W9-AXH-775

AN INTRODUCTORY GUIDE TO FINITE ELEMENT ANALYSIS

An Introductory Guide to Finite Element Analysis

A. A. BECKER

The American Society of Mechanical Engineers, New York

First published 2004

Published in the United Kingdom by
Professional Engineering Publishing Limited.

This edition published in the United States by ASME Press,
Three Park Avenue, New York, NY 10016, USA

ISBN 0-7918-0205-1

ASME order number 802051

A CIP catalogue record for this book is available from the British Library.

Contents

Author's Preface

Finite element (FE) technology has become very well established as the main tool for engineering analysis. Since the early 1980s, engineering analysts have extensively used FE software in analysing many engineering problems, ranging from relatively simple elastic analysis to nonlinear deformations of biological structures and crash simulations. The affordability and versatility of FE software has helped to spread its popularity. FE software is now very user-friendly and the task of applying the software has become relatively simple, requiring little or no training. Unfortunately, this has also resulted in the use of FE software by many analysts who are not familiar with the fundamental concepts of FE theory.

The real challenge in using FE software is not in generating complex three-dimensional meshes, but in knowing how to translate a real-life problem into a computational model and how to assess the accuracy of the FE solutions. A certain amount of engineering judgement is also needed. This book is intended to help inexperienced FE users in addressing these issues through a number of practical examples of FE applications chosen to illustrate how practical problems can be analysed using FE software.

The book is suitable both for beginners and those seeking to strengthen their background knowledge of FE methods. Throughout the book, no previous knowledge of FE is assumed. The book should provide readers with an engineering or physics background with the theoretical knowledge needed to appreciate the degree of approximation inherent in FE analysis.

The book starts with simple one-dimensional elements and gradually increases the complexity of the elements to cover pin-jointed elements, continuum elements, beam, plate, and shell elements. For all element formulations, the same derivation steps are used throughout the relevant chapters. Although the book is mainly concerned with linear stress analysis problems, brief introductions to nonlinear and thermal problems are also included.

The book is based on my lecturing material for undergraduate and postgraduate engineering courses at Imperial College and the University of Nottingham and the annual short courses aimed at practising engineers that have been running over the last 14 years. The feedback from students and course delegates has been very useful in improving the contents.

I would like to acknowledge the patience and support of my wife, Jane, my daughter, Leila, and my son, Alistair, who have been very understanding during the time-consuming task of writing this book. To them, the book is dedicated.

Professor A. A. Becker
Nottingham, UK

Notation

A_e	Element cross-sectional area
c_p	Specific heat
E	Young's modulus
F	Force
F^*	Force along a pin-jointed element
G	Number of Gaussian points
I	Second moment of area
J	Jacobian of transformation
K	Stiffness matrix
K_e	Element stiffness matrix
k	Thermal conductivity
L_e	Element length
M	Bending moment
N	Shape function
PDE	Partial Differential Equation
q	Rate of heat flow
S	Shear force
T	Temperature
t	Time
TPE	Total Potential Energy
U	Strain energy
u	Displacement
u^*	Displacement along a pin-jointed element
u_e	Element displacement
u_n	Normal displacement
u_{t}	Tangential displacement
W	Work done
w_g	Gaussian weight function
x^*	Coordinate along a pin-jointed element
x_g	Gaussian coordinate
α	Coefficient of thermal expansion
ε	Strain
$\varepsilon_{\mathrm{SB}}$	Stefan–Boltzmann constant

μ	Shear modulus
ν	Poisson's ratio
θ	Angle of inclination/slope
ρ	Density
σ	Stress
σ_e	Element stress

CHAPTER 1

Introduction and Background

Most numerical techniques in continuum mechanics are based on the principle that it is possible to derive some equations and relationships that accurately describe the behaviour of a small part of the body. By dividing the entire body into a large number of these smaller 'parts' or 'elements' and using appropriate compatibility and equilibrium relationships to link up or assemble these elements, it is possible to obtain a reasonably accurate prediction of the values of variables such as stresses and displacements in the body. As the sizes of these small elements are made smaller, the numerical solution becomes more accurate, but at the cost of increased computation time.

This chapter covers the theoretical background and notation used in this book. It deals with basic solid mechanics relationships, theories relevant to Finite Element (FE) concepts, and basic matrix algebra. The chapter serves as a refresher course for readers such as practising engineers who may not have used solid mechanics theory and notation recently.

1.1 NUMERICAL METHODS IN CONTINUUM MECHANICS

Numerical methods in continuum mechanics have traditionally been classified into three main approaches: finite element (FE), boundary element (BE), and finite difference (FD) methods (Fig. 1.1). The main features of these approaches are summarized below.

1.1.1 Finite element (FE) method

The main features of FE methods are:

- The entire solution domain is divided into small finite segments (hence the name 'finite elements').
- Over each element, the behaviour is described by the displacements of the elements and the material law.
- All elements are assembled together and the requirements of continuity and equilibrium are satisfied between neighbouring elements.
- Provided that the boundary conditions of the actual problem are satisfied, a unique solution can be obtained to the overall system of linear algebraic equations.

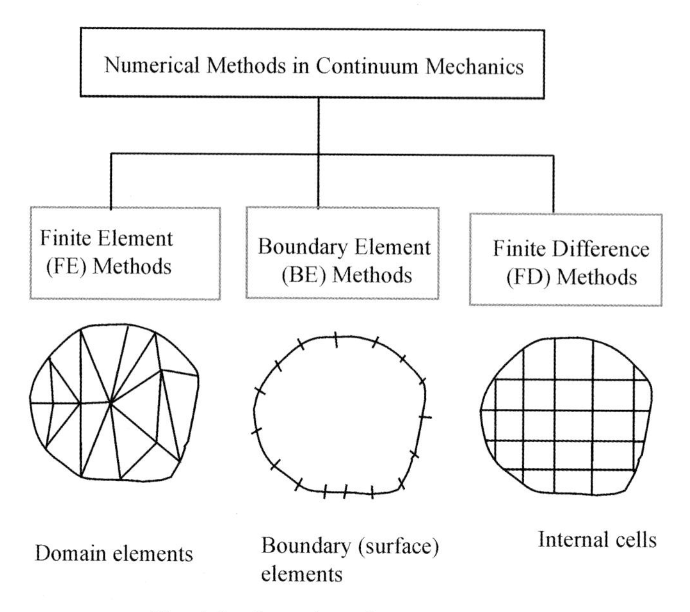

Fig. 1.1 Overview of numerical methods

- The solution matrix is sparsely populated (i.e. with relatively few non-zero coefficients).
- The FE method is very suitable for practical engineering problems of complex geometries. To obtain good accuracy in regions of rapidly changing variables, a large number of small elements must be used.

1.1.2 Boundary element (BE) method

The main features of BE methods are:

- The boundary (surface) is divided into small boundary segments ('boundary elements').
- The governing differential equations are transformed from volume integrals into surface or boundary integrals.
- The surface integrals are numerically integrated over each element.
- Provided that the boundary conditions of the actual problem are satisfied, a unique solution can be obtained to the overall system of linear algebraic equations.
- The solution matrix is fully populated.
- BE solutions at internal points are optional (rather than compulsory as in FE methods).

- The BE method can easily accommodate geometrically complex boundaries. Furthermore, because all the approximations are restricted to the surface, it can model regions with rapidly changing variables with better accuracy than the FE method.

Further information on boundary element methods can be found in several textbooks (**1–3**).

1.1.3 Finite difference (FD) method

The main features of FD methods are:

- The entire solution domain is divided into a grid of 'cells'.
- The derivatives in the governing partial differential equations are written in terms of finite difference equations.
- A finite difference approximation is applied to each interior point so that the displacement of each node is related to the values at the other nodes in the grid connected to it.
- Provided that the boundary conditions of the actual problem are satisfied, a unique solution can be obtained to the overall system of linear algebraic equations.
- The solution matrix is banded.
- The FD method is relatively easy to program. Its main serious drawback in practical engineering problems is that it is not suitable for problems with awkward irregular geometries. Furthermore, because it is difficult to vary the size of the difference cells in particular regions, it is not suitable for problems with rapidly changing variables such as stress concentration problems.
- FD methods are popular for heat transfer and fluid flow problems, rather than stress analysis problems.

Further information on finite difference methods can be found in many textbooks (**4**, **5**).

The above three approaches to modelling continuum mechanics problems are capable of arriving at accurate solutions provided a fine division of elements or internal cells is used. The FE and BE methods are well established in the analysis of solid mechanics.

1.2 Definition of stress

Stress is generally defined as the average force (F) per unit area (A). This definition assumes that the stress is uniform over that particular area, but in reality stresses are seldom uniform over large areas. Therefore, it is more meaningful if this area is made very small, thus introducing the mathematical concept of 'stress at a point', which is defined as

$$\sigma = \lim_{\delta A \to 0} \left(\frac{\delta F}{\delta A} \right) \tag{1.1}$$

The concept of stress at a point is physically valid because a small area δA would carry a small amount of force δF.

In a three-dimensional Cartesian axes system there are six components of stress (Fig. 1.2):

- three direct (tensile or compressive) stresses (σ_{xx}, σ_{yy}, σ_{zz}) caused by forces normal to the area; and
- three shear stresses (σ_{xy}, σ_{xz}, σ_{yz}) caused by shear forces acting parallel to the area.

The first subscript refers to the direction of the outward normal to the plane on which the stress acts, and the second subscript refers to the direction of the stress arrow. For simplicity, in most problems the first and second subscripts can be interchanged; in other words, $\sigma_{xy} = \sigma_{yx}$, $\sigma_{yz} = \sigma_{zy}$, and $\sigma_{xz} = \sigma_{zx}$ (complimentary shear stress).

A 'stress matrix' or a 'stress vector', which contains all stress components, can be conveniently expressed as

$$[\sigma] = \begin{bmatrix} \sigma_{xx} \\ \sigma_{yy} \\ \sigma_{zz} \\ \sigma_{xy} \\ \sigma_{xz} \\ \sigma_{yz} \end{bmatrix} \tag{1.2}$$

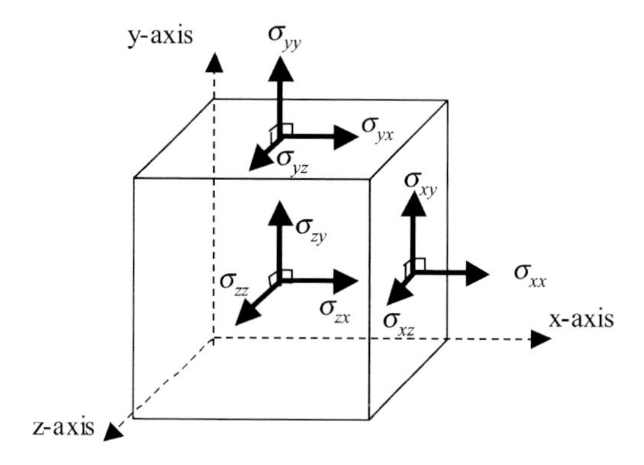

Fig. 1.2 Three-dimensional Cartesian stresses

Similarly, a 'strain vector' can be defined as

$$[\varepsilon] = \begin{bmatrix} \varepsilon_{xx} \\ \varepsilon_{yy} \\ \varepsilon_{zz} \\ \varepsilon_{xy} \\ \varepsilon_{xz} \\ \varepsilon_{yz} \end{bmatrix} \tag{1.3}$$

1.3 STRESS–STRAIN RELATIONSHIPS (HOOKE'S LAW)

Stress–strain relationships are often called 'constitutive equations'. For isotropic linear elastic materials with thermal strain, the following three-dimensional stress–strain equations (Hooke's law) can be used

$$\begin{aligned}
\varepsilon_{xx} &= \frac{1}{E}[\sigma_{xx} - \nu(\sigma_{yy} + \sigma_{zz})] + \alpha(\Delta T) \\
\varepsilon_{yy} &= \frac{1}{E}[\sigma_{yy} - \nu(\sigma_{xx} + \sigma_{zz})] + \alpha(\Delta T) \\
\varepsilon_{zz} &= \frac{1}{E}[\sigma_{zz} - \nu(\sigma_{xx} + \sigma_{yy})] + \alpha(\Delta T) \\
\varepsilon_{xy} &= \frac{1}{2\mu}\sigma_{xy} \\
\varepsilon_{xz} &= \frac{1}{2\mu}\sigma_{xz} \\
\varepsilon_{yz} &= \frac{1}{2\mu}\sigma_{yz}
\end{aligned} \tag{1.4}$$

where E is Young's modulus (N/m^2), ν is Poisson's ratio (dimensionless), μ is the shear modulus (N/m^2), α is the coefficient of thermal expansion (per °C), and ΔT is the temperature change from a reference value (°C).

The shear modulus μ is defined as

$$\mu = \frac{E}{2(1+\nu)} \tag{1.5}$$

Equation (1.4) can be rearranged such that the stresses are on the left-hand side, resulting in the following matrix expression, often referred to as the 'material constitutive equation'

$$[\sigma] = [D][\varepsilon] \tag{1.6}$$

where $[D]$ is called the 'elastic property matrix'.

1.4 STRAIN–DISPLACEMENT DEFINITIONS

The notation and sign convention used for strains are the same as those used for stresses. The three-dimensional direct (nonshear) strains are related to the displacements in the following relationships

$$\begin{aligned}\varepsilon_{xx} &= \frac{\partial u_x}{\partial x}\\ \varepsilon_{yy} &= \frac{\partial u_y}{\partial y}\\ \varepsilon_{zz} &= \frac{\partial u_z}{\partial z}\end{aligned} \tag{1.7}$$

where u_x, u_y, and u_z are the displacements (deformations) in the x, y, and z directions, respectively.

The three-dimensional shear strains are defined as

$$\begin{aligned}\varepsilon_{xy} &= \frac{1}{2}\left(\frac{\partial u_x}{\partial y} + \frac{\partial u_y}{\partial x}\right)\\ \varepsilon_{xz} &= \frac{1}{2}\left(\frac{\partial u_x}{\partial z} + \frac{\partial u_z}{\partial x}\right)\\ \varepsilon_{yz} &= \frac{1}{2}\left(\frac{\partial u_y}{\partial z} + \frac{\partial u_z}{\partial y}\right)\end{aligned} \tag{1.8}$$

It should be noted that two definitions are often used for the shear strain, one with the $\frac{1}{2}$ factor and one without. The shear strain definition with the $\frac{1}{2}$ factor (often referred to as the 'mathematical shear strain') is mainly used for the convenience of use in tensor notations. The shear strain definition without the $\frac{1}{2}$ factor is referred to as the 'engineering shear strain'. Both definitions are valid, provided that the definition is followed throughout the derivation of other relationships involving strains.

1.5 EQUILIBRIUM EQUATIONS

In order to examine the variation of stress from point to point within the body, consider the equilibrium of forces (resulting from the stresses) in the x, y, and z directions on a small differential volume. Figure 1.3 shows a small volume of dimensions δx, δy, and δz, where stresses change by a small amount. For example, σ_{xx} on one face changes to $(\sigma_{xx} + \delta\sigma_{xx})$ on the

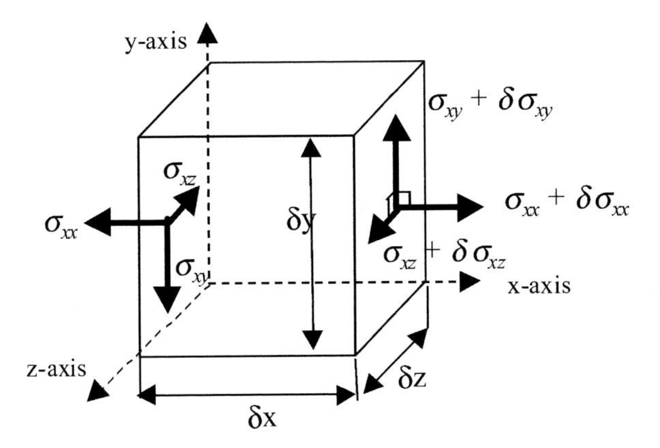

Fig. 1.3 Variation in stresses over a small differential volume

opposite face. Similarly, all other stress changes can be obtained. Therefore, the net force caused by the change in σ_{xx} in the x direction is given by

$$\delta\sigma_{xx}(\delta y\delta z) = \frac{\partial\sigma_{xx}}{\partial x}\delta x(\delta y\delta z) \tag{1.9}$$

Similarly, the other forces caused by stresses in the x direction are

$$\frac{\partial\sigma_{xy}}{\partial y}\delta y(\delta x\delta z) \qquad \frac{\partial\sigma_{xz}}{\partial z}\delta z(\delta x\delta y) \tag{1.10}$$

In addition to the net force in the x direction, there may be other forces caused by body forces (such as gravitational or centrifugal forces). If F_x is the net body force per unit volume acting in the x direction, then the equilibrium equation becomes

$$\delta x\delta y\delta z\left(\frac{\partial\sigma_{xx}}{\partial x} + \frac{\partial\sigma_{xy}}{\partial y} + \frac{\partial\sigma_{xz}}{\partial z}\right) + \delta x\delta y\delta z(F_x) = 0 \tag{1.11}$$

Similarly, all other net forces can be obtained, resulting in the following differential equations of stress known as the 'equilibrium equations'

$$\begin{aligned}
&\frac{\partial\sigma_{xx}}{\partial x} + \frac{\partial\sigma_{xy}}{\partial y} + \frac{\partial\sigma_{xz}}{\partial z} + F_x = 0 \\
&\frac{\partial\sigma_{yx}}{\partial x} + \frac{\partial\sigma_{yy}}{\partial y} + \frac{\partial\sigma_{yz}}{\partial z} + F_y = 0 \\
&\frac{\partial\sigma_{zx}}{\partial x} + \frac{\partial\sigma_{zy}}{\partial y} + \frac{\partial\sigma_{zz}}{\partial z} + F_z = 0
\end{aligned} \tag{1.12}$$

where F_x, F_y, and F_z are the body force components in the x, y, and z directions, respectively.

1.6 Compatibility Equations

In order to examine the strain variation from point to point within the body, differential equations of strains can be derived in a similar manner to the differential equations of stress (the equilibrium equations). These strain differential equations are called 'compatibility equations', and they must satisfy the physical constraint that the body must remain continuous as it deforms (i.e. no holes or overlaps created) as shown in Fig. 1.4. In other words, all displacements must be continuous and differentiable functions of position.

By eliminating u_x, u_y, and u_z using the strain–displacement definitions in equations (1.7) and (1.8) and differentiating with respect to the x, y, and z directions, the following six compatibility equations can be obtained (**6**)

$$
\begin{aligned}
\frac{\partial^2 \varepsilon_{xx}}{\partial y^2} + \frac{\partial^2 \varepsilon_{yy}}{\partial x^2} &= 2\frac{\partial^2 \varepsilon_{xy}}{\partial x \partial y} \\
\frac{\partial^2 \varepsilon_{yy}}{\partial z^2} + \frac{\partial^2 \varepsilon_{zz}}{\partial y^2} &= 2\frac{\partial^2 \varepsilon_{yz}}{\partial y \partial z} \\
\frac{\partial^2 \varepsilon_{zz}}{\partial x^2} + \frac{\partial^2 \varepsilon_{xx}}{\partial z^2} &= 2\frac{\partial^2 \varepsilon_{zx}}{\partial z \partial x} \\
\frac{\partial^2 \varepsilon_{xx}}{\partial y \partial z} + \frac{\partial^2 \varepsilon_{yz}}{\partial x^2} &= \frac{\partial^2 \varepsilon_{zx}}{\partial x \partial y} + \frac{\partial^2 \varepsilon_{xy}}{\partial x \partial z} \\
\frac{\partial^2 \varepsilon_{yy}}{\partial x \partial z} + \frac{\partial^2 \varepsilon_{zx}}{\partial^2 y} &= \frac{\partial^2 \varepsilon_{yz}}{\partial x \partial y} + \frac{\partial^2 \varepsilon_{xy}}{\partial y \partial z} \\
\frac{\partial^2 \varepsilon_{zz}}{\partial x \partial y} + \frac{\partial^2 \varepsilon_{xy}}{\partial z^2} &= \frac{\partial^2 \varepsilon_{yz}}{\partial x \partial z} + \frac{\partial^2 \varepsilon_{zx}}{\partial y \partial z}
\end{aligned}
\tag{1.13}
$$

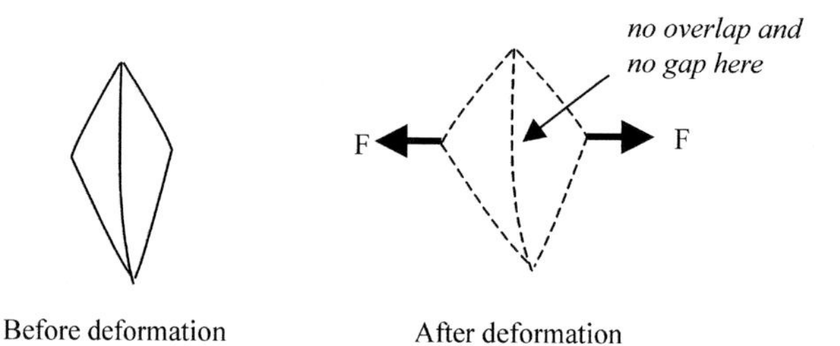

Fig. 1.4 Compatibility requirements between adjacent elements

If displacements are used as the independent variables in the FE formulation, then the compatibility equations are automatically satisfied. Hence these compatibility equations will not be explicitly used hereafter. If other variables such as stresses are used as the independent variables, then stress functions must be chosen to explicitly satisfy the above compatibility equations. This is the main reason why displacements are chosen as the independent variables in most FE formulation.

1.7 PRINCIPLE OF MINIMUM TOTAL POTENTIAL ENERGY

1.7.1 Stable and unstable problems

Static equilibrium can be stable, unstable, or neutral. For example, a ball inside a concave surface is in stable equilibrium because, if displaced sideways, it will try to return to its rest position, that is, the system reaches a position such that its potential energy is minimum. On the other hand, a ball on a convex surface will not return to its position if displaced sideways, and the potential energy is maximum. Neutral equilibrium is achieved when the ball is placed on a frictionless flat surface, that is, if displaced sideways it will stay in the new position and will not return to its old position. Here, the potential energy is unchanged with the displacement. Figure 1.5 summarizes the three types of static equilibrium.

In linear FE or structural analysis, the objective of the analysis is always to find a solution in which equilibrium is stable. The principle of minimum total potential energy (*TPE*) is often used in FE formulations by expressing the problem in terms of the independent variables (usually displacements) and then minimizing the *TPE* with respect to the displacements.

1.7.2 Strain energy

The work done by external forces on the body is stored in the form of strain energy. This strain energy is released upon the removal of the applied loads and the body returns to its undeformed state.

For linear elastic behaviour, an expression for the elastic strain energy can be derived as

$$U = \frac{1}{2}[\sigma][\varepsilon] \times volume \tag{1.14}$$

If the material behaviour is nonlinear, a more general expression can be written as

$$U = \int_{V}\int_{\varepsilon} \sigma \, \mathrm{d}\varepsilon \, \mathrm{d}V \tag{1.15}$$

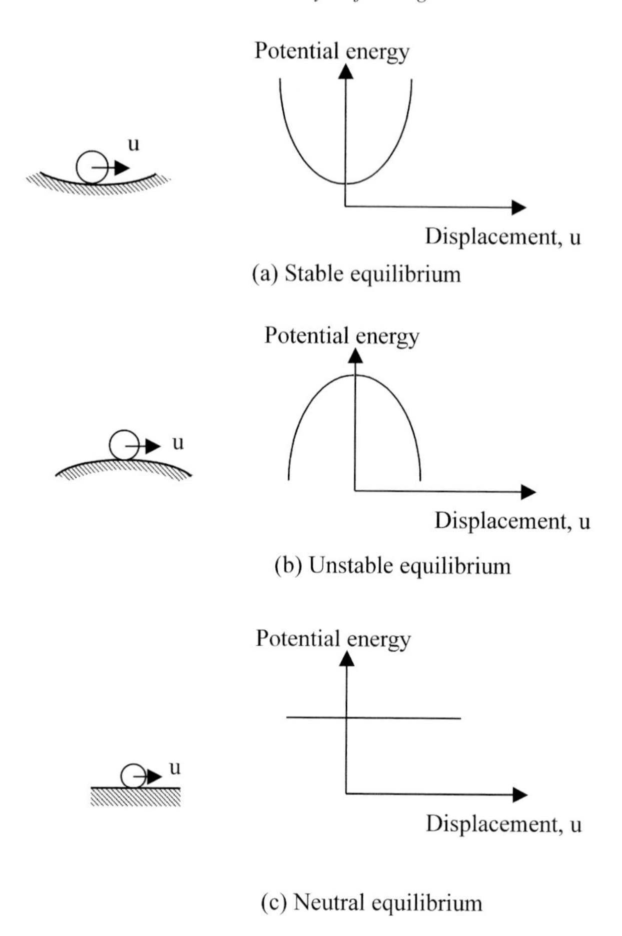

Fig. 1.5 Static equilibrium states

1.7.3 Work done by external forces

Another form of potential energy arises from the work done by the external forces that cause deformation of the body. This energy can be written as

$$W = \sum_i F_i u_i \tag{1.16}$$

where i is any point where the force F_i causes a displacement u_i. More general expressions can be derived for distributed loads or body forces such as centrifugal or gravitational loads.

1.7.4 Total potential energy (TPE)

The total potential energy (*TPE*) can be expressed as the difference between the strain energy and the work done by the external forces, that is

$$TPE = U - W \tag{1.17}$$

The principle of minimum total potential energy states that when the body is in equilibrium, the value of *TPE* must be stationary with respect to the variables of the problem. The equilibrium is stable if the *TPE* is minimum (and unstable if the *TPE* is maximum).

In most FE formulations, the displacement, u, is chosen as the unknown (i.e. independent) variable of the problem, and the principle of minimum *TPE* requires that the *TPE* is stationary, such that

$$\frac{\partial(TPE)}{\partial u} = 0 \tag{1.18}$$

In FE formulations, it can be shown that the stationary *TPE* is minimum. This minimization of the *TPE* can be carried out for each element in a finite element model, assuming that the interfaces between the elements make no contribution to the *TPE*.

1.8 Matrix definitions

Matrices are often used to simplify writing lengthy algebraic equations. For example, the three equations

$$\begin{aligned} A_{11}x_1 + A_{12}x_2 + A_{13}x_3 &= b_1 \\ A_{21}x_1 + A_{22}x_2 + A_{23}x_3 &= b_2 \\ A_{31}x_1 + A_{32}x_2 + A_{33}x_3 &= b_3 \end{aligned} \tag{1.19}$$

can be expressed as the following matrices

$$\begin{bmatrix} A_{11} & A_{12} & A_{13} \\ A_{21} & A_{22} & A_{23} \\ A_{31} & A_{32} & A_{33} \end{bmatrix} \begin{bmatrix} x_1 \\ x_2 \\ x_3 \end{bmatrix} = \begin{bmatrix} b_1 \\ b_2 \\ b_3 \end{bmatrix} \tag{1.20}$$

or, in a more concise form

$$[A][x] = [b] \tag{1.21}$$

where $[A]$ is a 3×3 matrix, $[x]$ and $[b]$ are 3×1 matrices. In general, equation (1.21) can be used to represent any number of equations. If N is the total number of equations, then $[A]$ is an $N \times N$ matrix, and $[x]$ and $[b]$ are $N \times 1$ matrices. Note that matrices such as $[x]$ and $[b]$ with just one column are sometimes called 'vectors'.

1.8.1 Matrix multiplication

By observation of equations (1.19) and (1.20), the matrix multiplication rules can be easily learned. In general, if two matrices $[A]$ and $[B]$ are multiplied, then the number of columns of $[A]$ must be the same as the number of rows of $[B]$; that is, if $[A]$ is an $(m \times n)$ matrix, and $[B]$ is a $(p \times q)$ matrix, then n must be equal to p. The resulting matrix $[C]$ is an $(m \times q)$ matrix.

$$[A]^{(m\times n)} \times [B]^{(p\times q)} = [C]^{(m\times q)} \qquad (n \text{ must be equal to } p) \tag{1.22}$$

Note that, in general, $[A] \times [B]$ is not equal to $[B] \times [A]$.

1.8.2 Transpose of a matrix

If the rows and columns of a matrix $[A]$ are interchanged, the resulting matrix is called the transpose of $[A]$ or $[A]^T$. If the elements of $[A]$ are written as a_{ij}, then the elements of $[A]^T$ are a_{ji}. The following example shows a matrix $[A]$ and its transpose $[A]^T$.

$$[A] = \begin{bmatrix} 4 & 2 & 6 & 9 \\ 3 & 7 & 8 & 2 \\ 17 & 5 & 5 & 11 \\ 22 & 7 & 8 & 1 \end{bmatrix}; \qquad [A]^T = \begin{bmatrix} 4 & 3 & 17 & 22 \\ 2 & 7 & 5 & 7 \\ 6 & 8 & 5 & 8 \\ 9 & 2 & 11 & 1 \end{bmatrix} \tag{1.23}$$

The following relationships are useful.

$$\begin{aligned} ([A] \times [B])^T &= [B]^T \times [A]^T \\ ([A]^T)^T &= [A] \end{aligned} \tag{1.24}$$

1.8.3 Symmetric matrix

A square matrix (number of rows equal to the number of columns) is called symmetric if $[A]^T = [A]$, that is, $a_{ij} = a_{ji}$. This means that matrix coefficients above the diagonal of the matrix are 'mirror images' of those below the diagonal. For example, the following square (4×4) matrix is symmetric.

$$[A] = \begin{bmatrix} 4 & 2 & 6 & 9 \\ 2 & 7 & 8 & 2 \\ 6 & 8 & 5 & 7 \\ 9 & 2 & 7 & 1 \end{bmatrix} \tag{1.25}$$

In FE formulations, the stiffness matrices are symmetrical, and it is important to exploit this symmetry to economize on the storage requirements of large matrices.

1.8.4 Inverse of a matrix

A 'unit matrix', $[I]$, is a square matrix in which all the coefficients of the principal diagonal are equal to 1, while all other coefficients are zero. In other words

$$[I] = \begin{bmatrix} 1 & 0 & 0 & 0 \\ 0 & 1 & 0 & 0 \\ 0 & 0 & 1 & 0 \\ 0 & 0 & 0 & 1 \end{bmatrix} \tag{1.26}$$

If for a given matrix $[A]$ there exists a matrix $[B]$ such that $[A]\,[B] = [I]$, where $[I]$ is a 'unit matrix', then $[B]$ is called the 'inverse' of $[A]$ and is denoted by $[A^{-1}]$ such that

$$[A] \times [A^{-1}] = [I] \tag{1.27}$$

Therefore, to solve a system of linear algebraic equations, such as that shown in equation (1.21), both sides of the equation can be multiplied by $[A^{-1}]$ to give

$$[x] = [A^{-1}][b] \tag{1.28}$$

However, for large matrices, such as those associated with FE analysis, it is important to use efficient numerical methods for solving large systems of algebraic equations. Inverting a matrix requires a substantial number of mathematical operations, for example, of the order N^4 where N is the number of equations. In practice, the inverse of $[A]$ is not directly computed. Instead, special equation-solving methods such as 'Gaussian elimination' or iterative 'Gauss–Seidel' techniques are used (**7**).

1.8.5 Strain energy matrix example

To demonstrate the advantages of using matrices in FE formulations, consider a one-dimensional (uniaxial) problem where the strain energy stored in the body, per unit volume, is given by

$$U = \frac{1}{2}\sigma_{xx}\varepsilon_{xx} \tag{1.29}$$

This expression can be generalized for two-dimensional problems as

$$U = \frac{1}{2}(\sigma_{xx}\varepsilon_{xx} + \sigma_{yy}\varepsilon_{yy} + \sigma_{xy}\varepsilon_{xy}) \tag{1.30}$$

Similarly, for three-dimensional problems

$$U = \frac{1}{2}(\sigma_{xx}\varepsilon_{xx} + \sigma_{yy}\varepsilon_{yy} + \sigma_{zz}\varepsilon_{zz} + \sigma_{xy}\varepsilon_{xy} + \sigma_{xz}\varepsilon_{xz} + \sigma_{yz}\varepsilon_{yz}) \tag{1.31}$$

Rather than dealing with three equations, it is much more convenient to express all the above equations in a single equation applicable to any dimensionality. Using matrices, all three equations can be combined as

$$U = \frac{1}{2}[\sigma]^T[\varepsilon] \tag{1.32}$$

where $[\sigma]$ and $[\varepsilon]$ are the stress and strain vectors, defined in equations (1.2) and (1.3), respectively.

Note that $[\sigma]^T$, that is, the transpose of the stress matrix, is used in equation (1.32) in order to satisfy the matrix multiplication rules. Because $[\sigma]^T$ is a 1×6 matrix and $[\varepsilon]$ is a 6×1 matrix, their product is a 1×1 matrix, that is, a matrix with only one coefficient.

Alternatively, it is possible to use a 'tensor' notation

$$U = \frac{1}{2}\sigma_{ij}\varepsilon_{ij} \tag{1.33}$$

where the subscripts i and j take the values 1, 2, and 3 corresponding to the Cartesian directions x, y, and z, respectively.

Both matrix and tensor notations are widely used in FE formulations. In this book, only matrix expressions will be used hereafter.

1.9 Layout of the Book

This chapter has covered the relevant theoretical background and matrix notation used in this book. The rest of the book is divided into the following chapters.

Chapter 2 (*Structural Analysis Using Pin-Jointed Elements*) presents a very simplified approach to the derivation of simple FE formulations for pin-jointed structures, avoiding the use of complex theorems. A solved example is presented to demonstrate how the FE matrices are assembled.

Chapter 3 (*Continuum Elements*) builds on the procedures introduced in the simple structural analysis, in order to derive a formulation for two-dimensional, axisymmetric, and three-dimensional continuum elements.

Chapter 4 (*Energy and Variational Approaches*) presents details of the energy and other theorems used in the derivation of modern FE formulations. Although mathematical derivations are presented in this chapter, emphasis is placed on the physical significance of these theories and the degree of approximation associated with them. Two illustrative examples of a cantilever beam and a flexible cable are used to demonstrate the application of these theorems.

Chapter 5 (*Higher Order Quadratic Elements*) covers more advanced formulations for more sophisticated quadratic elements. The FE concepts of isoparametric elements, shape functions, local and global coordinates, and numerical integrations are presented.

Chapter 6 (*Beam, Plate and Shell Elements*) presents the formulations of beam, plate, and shell elements, and emphasizes the different degrees of freedom used in these elements.

Chapter 7 (*Practical Guidelines for FE Applications*) is dedicated to the application of FE software to practical problems. It presents a number of guidelines, without the use of mathematical equations, on how to adopt good practice in applying FE software.

Chapter 8 (*Introduction to Nonlinear FE Analysis*) introduces the reader to more advanced nonlinear FE applications. This is intended to enable the reader to differentiate between linear and nonlinear problems, and to highlight the main difficulties associated with nonlinear FE analysis.

Chapter 9 (*Thermal Problems*) aims to demonstrate how the FE formulation can be applied to a nonsolid mechanics problem. A theoretical background to thermal problems is given, followed by the FE formulation based on simple elements. Transient problems are also presented to highlight how time-marching is incorporated in FE analysis.

Chapter 10 (*Examples of FE Applications*) presents a number of practical FE examples. Each problem is analysed from an FE viewpoint to highlight the best approach to obtain a good FE solution for the stresses and displacements. Various aspects of FE analysis are illustrated, such as choosing the correct type of element, effect of mesh refinement, preventing rigid body motion, and simplifying geometric features.

A comprehensive glossary is included to provide an explanation of all the technical terminology used in FE analysis.

CHAPTER 2

Structural Analysis Using Pin-Jointed Elements

Finite element methods can be introduced as a numerical procedure for analysing structural pin-jointed problems. In this chapter, FE analysis of a simple one-dimensional tension element is first described and then extended to more general pin-jointed structures using 'truss' elements. These structures consist of long thin elements linked together by frictionless pin-joints, which are assumed to transmit only axial forces to the elements. The analysis shown here is confined to plane (two-dimensional) frames, but can be extended to three-dimensional space frames. Similarly, although the formulation presented here is for two-node linear elements, it can be extended to more sophisticated elements, with more nodes per element.

The main objective of the FE analysis of pin-jointed structures is to determine both the forces in the elements and the displacements of the joints (called nodes), under specified conditions of loading. The basic strategy of most FE formulations is to treat the nodal displacements as the unknown (independent) variables to be determined by solving a system of linear algebraic equations.

2.1 A SIMPLE ONE-DIMENSIONAL TENSION ELEMENT

A bar element subjected to an axial load is considered the most simple type of element. Figure 2.1 shows a straight axial bar of length L_e and cross-sectional area A_e, with two points at either end (called 'nodes'). The deformations (displacements) of nodes 1 and 2 are u_1 and u_2, respectively, with corresponding forces F_1 and F_2.

Assuming that the material is linear elastic, the stress–strain relationships are given by

$$\sigma = E\varepsilon \tag{2.1}$$

Strain is defined as the change in length divided by the original length

$$\varepsilon = \frac{\Delta L}{L_e} = \frac{u_1 - u_2}{L_e} \tag{2.2}$$

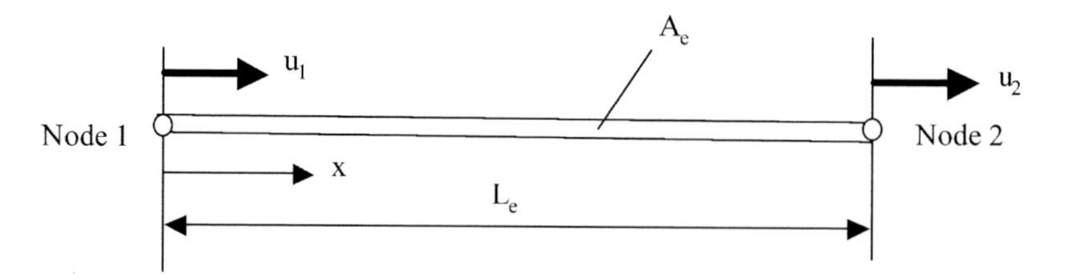

Fig. 2.1 A one-dimensional bar element

The stress in the bar is given by

$$\sigma = \frac{F}{A} \tag{2.3}$$

Substituting the stress and strain in equation (2.1) results in

$$\frac{F}{AE} = \frac{u_1 - u_2}{L_e} \tag{2.4}$$

Hence a force–displacement relationship can be obtained as

$$F = \frac{AE}{L_e}(u_1 - u_2) \tag{2.5}$$

At nodes 1 and 2, the forces are obtained as

$$\begin{aligned} \frac{AE}{L_e}(u_1 - u_2) &= F_1 \\ \frac{AE}{L_e}(u_2 - u_1) &= F_2 \end{aligned} \tag{2.6}$$

which can be expressed in matrix form as

$$\left(\frac{EA}{L_e}\right)\begin{bmatrix} 1 & -1 \\ -1 & 1 \end{bmatrix}\begin{bmatrix} u_1 \\ u_2 \end{bmatrix} = \begin{bmatrix} F_1 \\ F_2 \end{bmatrix} \tag{2.7}$$

or, in a more concise way

$$[k_e][u_e] = [F_e] \tag{2.8}$$

where $[k_e]$ is the 'element stiffness matrix' (similar to a spring stiffness), and $[u_e]$ and $[F_e]$ are the element displacement and force vectors, respectively.

This simple bar can only carry loads in the axial direction, and is thus a one-dimensional element, which cannot be used for problems with bending moments or shear loads. Such an element is said to have one degree of freedom, that is, only one variable, u.

Before moving to more sophisticated elements, such as truss elements used in pin-jointed structures, it is worth considering an alternative derivation of the element stiffness matrix based on an energy approach. The strain energy, U, per unit volume is given by

$$U = \frac{1}{2}[\sigma][\varepsilon] \tag{2.9}$$

Therefore, for an element of length L_e, the strain energy is

$$U = \int_0^{L_e} \frac{1}{2}[\sigma][\varepsilon]A\,\mathrm{d}x \tag{2.10}$$

Substituting for stress in terms of strain and using the strain–displacement definition results in

$$U = \int_0^{L_e} \frac{1}{2}E[\varepsilon]^2 A\,\mathrm{d}x = \int_0^{L_e} \frac{1}{2}E\left(\frac{\mathrm{d}u}{\mathrm{d}x}\right)^2 A\,\mathrm{d}x = \frac{EA}{2}\int_0^{L_e}\left(\frac{\mathrm{d}u}{\mathrm{d}x}\right)^2 \mathrm{d}x \tag{2.11}$$

For a two-node element, a linear displacement function is assumed as

$$u(x) = C_1 + C_2 x \tag{2.12}$$

where C_1 and C_2 are constants. This is called the 'trial function', which must satisfy the boundary conditions, that is, the displacement conditions at the two nodes:

- at node 1 (where $x = 0$), $u = u_1$;
- at node 2 (where $x = L_e$), $u = u_2$.

The constants C_1 and C_2 can therefore be expressed in terms of u_1 and u_2 as

$$\begin{aligned} u_1 &= C_1 + 0 \Rightarrow C_1 = u_1 \\ u_2 &= C_1 + C_2 L_e \Rightarrow C_2 = \frac{u_2 - u_1}{L_e} \end{aligned} \tag{2.13}$$

The displacement function can therefore be expressed in terms of the nodal displacements as

$$u(x) = u_1 + \left(\frac{u_2 - u_1}{L_e}\right)x \tag{2.14}$$

which can be rearranged to give

$$u(x) = \left(1 - \frac{x}{L_e}\right)u_1 + \left(\frac{x}{L_e}\right)u_2 \tag{2.15}$$

This process is similar to curve-fitting, where a straight line equation is obtained from the coordinates of two points at either end. If more nodes are used per element, the expression for the displacement function becomes quadratic or higher order. The functions that multiply the nodal displacements u_1 and u_2 are called the 'shape functions' or the 'interpolation functions'. For the two-node element used here, the shape functions are linear.

Differentiating the displacement function gives

$$\frac{\mathrm{d}u}{\mathrm{d}x} = \left(\frac{-1}{L_e}\right)u_1 + \left(\frac{1}{L_e}\right)u_2 = \frac{u_2 - u_1}{L_e} \tag{2.16}$$

Hence, the strain energy expression becomes

$$\begin{aligned} U &= \frac{EA}{2}\int_0^{L_e}\left(\frac{u_2 - u_1}{L_e}\right)^2 \mathrm{d}x = \frac{EA}{2L_e^2}(u_2 - u_1)^2[x]_0^{L_e} \\ &= \frac{EA}{2L_e}(u_2^2 - 2u_2u_1 + u_1^2) \end{aligned} \tag{2.17}$$

The principle of minimum total potential energy (*TPE*), discussed in Section 1.7, can be used to minimize the strain energy function with respect to the nodal displacements. The principle states that

$$TPE = U - W \tag{2.18}$$

where W is the work done by the forces F_1 and F_2, such that

$$W = F_1u_1 + F_2u_2 \tag{2.19}$$

Therefore, the *TPE* expression can be written in terms of the displacements as

$$TPE = \frac{EA}{2L_e}(u_2^2 - 2u_2u_1 + u_1^2) - (F_1u_1 + F_2u_2) \tag{2.20}$$

Minimizing *TPE* with respect to u_1 gives

$$\frac{\partial TPE}{\partial u_1} = 0 = \frac{EA}{2L_e}(-2u_2 + 2u_1) - F_1 \tag{2.21}$$

Similarly, minimizing *TPE* with respect to u_2 gives

$$\frac{\partial TPE}{\partial u_2} = 0 = \frac{EA}{2L_e}(2u_2 - 2u_1) - F_2 \quad (2.22)$$

The above two equations can be combined in matrix form as

$$\left(\frac{EA}{L_e}\right)\begin{bmatrix} 1 & -1 \\ -1 & 1 \end{bmatrix}\begin{bmatrix} u_1 \\ u_2 \end{bmatrix} = \begin{bmatrix} F_1 \\ F_2 \end{bmatrix} \quad (2.23)$$

which is identical to equation (2.7).

2.2 PIN-JOINTED STRUCTURES (TRUSS ELEMENTS)

The derivation of the FE formulation of structural problems can be broken down into a number of steps, beginning with defining the element and interpolation functions, assembling the overall set of equations and then solving the equations to obtain all the variables. It is convenient to divide the FE formulation into seven steps as described below.

Step 1: Define the element and the shape (interpolation) functions
The first step in any FE analysis is to divide the structure into elements and examine the behaviour of a typical element. Consider a simple linear truss element, of length L_e and cross-sectional area A_e, with one node at either end, and oriented at an angle θ anticlockwise from the x-direction, as shown in Fig. 2.2. Note that a global origin from which all the x- and y-coordinates of the nodes are measured should be defined. For convenience, a local direction x^* along the element is also defined.

The element is assumed to have two degrees of freedom, the global (Cartesian) displacements in the x- and y-directions, u_x and u_y, which are components of the local displacement, u^*, along the element (in the direction of x^*). F_x and F_y are the global (Cartesian) forces in the x- and y-directions, which are components of the local force, F^*, along the element (in the direction of x^*).

For simplicity, the displacements, which are the unknown variables, will be assumed to vary linearly over each element, that is, a constant strain (and stress) within each element

$$u^* = C_1 + C_2x^* \quad (2.24)$$

where C_1 and C_2 are constants.

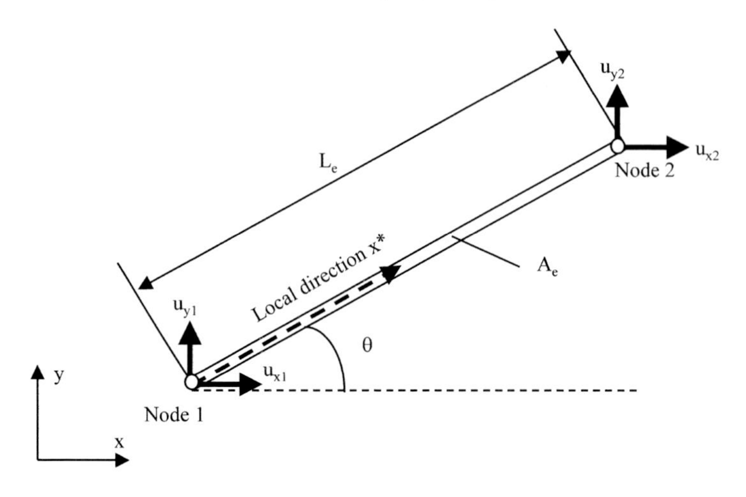

Fig. 2.2 A two-node truss element

The displacement conditions at the two nodes are:

- at node 1 (where $x^* = 0$), $u^* = u_1^*$;
- at node 2 (where $x^* = L_e$), $u^* = u_2^*$.

Therefore, the displacements of the two nodes can be written in terms of C_1 and C_2 in matrix form as

$$\begin{bmatrix} u_1^* \\ u_2^* \end{bmatrix} = \begin{bmatrix} 1 & 0 \\ 1 & L_e \end{bmatrix} \begin{bmatrix} C_1 \\ C_2 \end{bmatrix} \tag{2.25}$$

These equations can be generalized as

$$[u_e^*] = [A][C] \tag{2.26}$$

where $[u_e^*]$ is the displacement vector of the element, $[A]$ is a 'coordinate matrix', and $[C]$ contains the constants C_1 and C_2. It should be noted that the number of constants (here C_1 and C_2) must be equal to the number of nodal points; for example, if an element has three nodes, three constants should be used to describe a quadratic behaviour of displacement. The element described here has a linear 'shape function'. In order to determine the constants C_1 and C_2, the matrix $[C]$ can be expressed as

$$[C] = [A^{-1}][u_e^*] \tag{2.27}$$

Although it is convenient to use the displacement vector $[u_e^*]$ along the element, the global (Cartesian) x and y components of the displacement vector are often used in practice. It is also convenient to group relevant variables together as matrices. This simplifies the expressions, and makes them general enough to be applicable to other problems such as three-dimensional problems or more sophisticated elements. The following element matrices (vectors) can be defined.

$$[u_e] = \begin{bmatrix} u_{x1} \\ u_{y1} \\ u_{x2} \\ u_{y2} \end{bmatrix}; \qquad [F_e] = \begin{bmatrix} F_{x1} \\ F_{y1} \\ F_{x2} \\ F_{y2} \end{bmatrix} \tag{2.28}$$

where $[u_e]$ is the Cartesian element displacement vector and $[F_e]$ is the Cartesian element force vector. Note the order of the components in the vectors.

By resolving the displacements along the direction of the element, the global displacement components can be easily determined as

$$\begin{aligned} u_1^* &= u_{x1} \cos\theta + u_{y1} \sin\theta \\ u_2^* &= u_{x2} \cos\theta + u_{y2} \sin\theta \end{aligned} \tag{2.29}$$

Substituting these expressions in equation (2.27) results in the expression

$$[C] = \begin{bmatrix} C_1 \\ C_2 \end{bmatrix} = \begin{bmatrix} \cos\theta & \sin\theta & 0 & 0 \\ \dfrac{-\cos\theta}{L_e} & \dfrac{-\sin\theta}{L_e} & \dfrac{\cos\theta}{L_e} & \dfrac{\sin\theta}{L_e} \end{bmatrix} \begin{bmatrix} u_{x1} \\ u_{y1} \\ u_{x2} \\ u_{y2} \end{bmatrix} \tag{2.30}$$

Step 2: Satisfy the material law (constitutive equations)

To derive an expression for the element stiffness, that is, element forces as functions of element displacements, the displacements are differentiated (to obtain the strains), and then substituted in the material law, in other words, the stress–strain relationships.

Because the displacement variable u^* is along the element, only one strain component exists, which can be easily determined by differentiating equation (2.24), such that

$$[\varepsilon] = \frac{du^*}{dx^*} = \frac{d}{dx^*}(C_1 + C_2 x^*) = C_2 \tag{2.31}$$

Note that the strain per element is therefore constant. Substituting C_2 from equation (2.30) gives the following expression for strain in terms of the displacement vector $[u_e]$, such that

$$[\varepsilon] = \frac{1}{L_e}[-\cos\theta \quad -\sin\theta \quad \cos\theta \quad \sin\theta] \begin{bmatrix} u_{x1} \\ u_{y1} \\ u_{x2} \\ u_{y2} \end{bmatrix} \tag{2.32}$$

or, in general

$$[\varepsilon] = [B][u_e] \tag{2.33}$$

where $[B]$ is a 'dimension matrix', defined as

$$[B] = \frac{1}{L_e}[-\cos\theta \quad -\sin\theta \quad \cos\theta \quad \sin\theta] \tag{2.34}$$

The generalized Hooke's law expressions for strain–stress can be written in matrix form as

$$[\sigma] = [D][\varepsilon] \tag{2.35}$$

where $[D]$ is a 'material property matrix'. In the truss element, $[D] = E$, because the load is uniaxial.

Stresses can now be expressed as a function of displacements by substituting equation (2.33) such that

$$[\sigma] = [D][B][u_e] \tag{2.36}$$

Step 3: Derive the element stiffness matrix

Two alternative approaches can be used to derive the element stiffness matrix: a direct equilibrium of forces approach, and a more general energy approach.

Direct equilibrium approach

To satisfy equilibrium at the nodes, the forces at each node can be written as (Fig. 2.3)

$$\begin{aligned} F_{x1} &= -F^* \cos\theta \\ F_{y1} &= -F^* \sin\theta \\ F_{x2} &= F^* \cos\theta \\ F_{y2} &= F^* \sin\theta \end{aligned} \tag{2.37}$$

By using the expression for matrix $[B]$ from equation (2.34), these force expressions can be expressed in matrix form as

$$[F_e] = L_e[B]^T F^* \tag{2.38}$$

The element stress can be simply obtained by dividing the uniaxial force F^* by the element area, such that

$$F^* = A_e[\sigma] \tag{2.39}$$

Substituting equations (2.38) and (2.36) into equation (2.39), an expression for the element forces in terms of displacements can be written as

$$[F_e] = A_e L_e [B]^T [D][B][u_e] \tag{2.40}$$

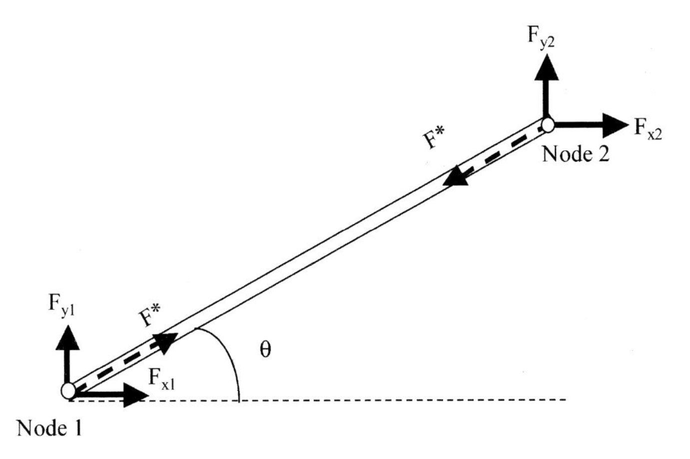

Fig. 2.3 Equilibrium of the nodal points

Therefore the element stiffness matrix $[k_e]$ can be defined as

$$[F_e] = [k_e][u_e] \tag{2.41}$$

where $[k_e]$ is expressed as

$$[k_e] = A_e L_e [B]^T [D][B] \tag{2.42}$$

By substituting for $[B]$ and $[D]$, and further manipulation, the element stiffness matrix can be expressed as

$$[k_e] = \left(\frac{A_e E}{L_e}\right) \begin{bmatrix} \cos^2\theta & \cos\theta\sin\theta & -\cos^2\theta & -\cos\theta\sin\theta \\ \cos\theta\sin\theta & \sin^2\theta & -\cos\theta\sin\theta & -\sin^2\theta \\ -\cos^2\theta & -\cos\theta\sin\theta & \cos^2\theta & \cos\theta\sin\theta \\ -\cos\theta\sin\theta & -\sin^2\theta & \cos\theta\sin\theta & \sin^2\theta \end{bmatrix} \tag{2.43}$$

Note that the element stiffness matrix is a symmetric matrix.

Energy method

Using an energy formulation to derive the stiffness matrix is more robust than the procedure adopted above using equilibrium conditions. This is because the energy formulation also applies to other types of elements (such as quadratic or cubic elements) and to three-dimensional problems.

The total potential energy (*TPE*) of the element can be expressed as

$$TPE = \int_v \frac{1}{2}[\sigma]^T[\varepsilon]\,\mathrm{d}v - [u_e]^T[F_e] \tag{2.44}$$

where ν is the volume (here $\mathrm{d}\nu = A_e\,\mathrm{d}x$). Substituting for the stress and strain from equations (2.36) and (2.33), respectively, results in

$$\begin{aligned} TPE &= \int_\nu \frac{1}{2}([D][B][u_e])^T([B][u_e])(A_e\,\mathrm{d}x) - [u_e]^T[F_e] \\ &= \int_0^{L_e} \frac{1}{2}A_e[u_e]^T\,[B]^T[D][B][u_e]\,\mathrm{d}x - [u_e]^T[F_e] \end{aligned} \tag{2.45}$$

Note that $[D]^T = [D]$ since $[D]$ is symmetric. Using the principle of minimum total potential energy, the differential of *TPE* with respect to the displacement $[u_e]$ must be zero, that is

$$\partial\frac{(TPE)}{\partial[u_e]} = 0 = A_e[B]^T[D][B][u_e]L_e - [F_e] \tag{2.46}$$

which leads to the following expression for the element stiffness

$$0 = [k_e][u_e] - [F_e] \tag{2.47}$$

which is identical to the stiffness matrix derived using the direct equilibrium approach. Note that in equation (2.46), differentiation with respect to the matrix $[u_e]$ can be performed in a similar manner to an algebraic differentiation, that is, the differential of $[u_e]^T[u_e]$ is similar to differentiating $[u_e]^2$ for which the differential is $2[u_e]$.

Step 4: Assemble the overall stiffness matrix

The next stage is to assemble all the individual elements to form the overall (global) structure. This is done either by using an energy formulation or using the equilibrium and compatibility conditions at the nodes. The displacement of a particular node must be the same for every element connected to it. The externally applied forces at the nodes must also be balanced by the forces on the elements at these nodes. In other words

$$[F]_{\text{external}} = \sum_{\text{elements}} [F_e] = \sum_{\text{elements}} [k_e][u_e] \tag{2.48}$$

Therefore, a global system of equations can be written as

$$[K]_{\text{global}}\,[u]_{\text{global}} = [F]_{\text{global}} \tag{2.49}$$

where the matrices shown are the global matrices containing all the nodal points. For a problem with N nodes with two degrees of freedom per node, the global $[K]$ is of size $(2N \times 2N)$, while the global $[u]$ and $[F]$ vectors are of size $(2N \times 1)$, as follows.

$$
\underset{(2N \times 2N)}{\begin{bmatrix} K_{11} & K_{12} & K_{13} & K_{14} & \cdot & \cdot & \cdot & K_{1,2N} \\ K_{21} & K_{22} & K_{23} & K_{24} & \cdot & \cdot & \cdot & K_{2,2N} \\ \cdot & \cdot & \cdot & \cdot & \cdot & \cdot & \cdot & \cdot \\ \cdot & \cdot & \cdot & \cdot & \cdot & \cdot & \cdot & \cdot \\ \cdot & \cdot & \cdot & \cdot & \cdot & \cdot & \cdot & \cdot \\ \cdot & \cdot & \cdot & \cdot & \cdot & \cdot & \cdot & \cdot \\ \cdot & \cdot & \cdot & \cdot & \cdot & \cdot & \cdot & \cdot \\ \cdot & \cdot & \cdot & \cdot & \cdot & \cdot & \cdot & K_{2N,2N} \end{bmatrix}}
\underset{(2N \times 1)}{\begin{bmatrix} u_{x1} \\ u_{y1} \\ u_{x2} \\ u_{y2} \\ \cdot \\ \cdot \\ u_{xN} \\ u_{yN} \end{bmatrix}}
=
\underset{(2N \times 1)}{\begin{bmatrix} F_{x1} \\ F_{y1} \\ F_{x2} \\ F_{y2} \\ \cdot \\ \cdot \\ F_{xN} \\ F_{yN} \end{bmatrix}}
\tag{2.50}
$$

Note that the global stiffness matrix $[K]$ is symmetric and sparsely populated (containing relatively few nonzero coefficients) in structures containing a large number of elements. This is because not more than a few elements are connected to any one node.

Step 5: Apply the boundary conditions and external loads
To obtain a unique solution of the problem, some displacement constraints (i.e. boundary conditions) and loading conditions must be prescribed at some of the nodes. These usually take one of the following forms:

(a) Displacement condition: A zero or nonzero prescribed nodal displacement or sliding at an angle;
(b) Force condition: A prescribed applied force in a given direction.

These boundary conditions can be incorporated into the system of linear algebraic equations, which can then be solved to obtain a unique solution for the displacements at each node.

Step 6: Solve the equations
Standard equation solvers, such as the Gaussian elimination technique, can be used to solve the equations to determine the unknown variables (here displacements) at each node. Because the stiffness matrix is symmetric and sparsely populated, specially adapted solvers that can considerably reduce the computational time and storage requirements are often used in FE codes.

Step 7: Compute other variables
After solving the global equations, displacements at all the nodal points are determined. From the displacement values, the element strains can be obtained from the strain–displacement relationship in equation (2.33), and

the element stresses from equation (2.36). Note that in most FE formulations, only the displacements are the independent variables, that is, forces, strains, and stresses are obtained from the computed displacements.

2.3 A STRUCTURAL ANALYSIS EXAMPLE

The following simple structural analysis example demonstrates how the global stiffness matrix $[K]$ is assembled from stiffness matrices of the individual elements, $[k_e]$.

Problem definition

A frictionless pin-jointed structure consisting of three members is shown in Fig. 2.4. Points 1, 2, and 4 are fixed to a rigid surface, and a point force of 20 kN is applied to point 3 at an angle of 45°. The members have either a cross-sectional area of A_1 (250 mm^2) or A_2 (450 mm^2). Young's modulus for all members is 207 GN/m^2.

The FE formulation developed above will be used to calculate the horizontal and vertical components of the displacement at point 3. The element stresses will be obtained from the calculated displacements.

Element stiffness matrix for element e1

The element stiffness, $[k_e]$, has been previously derived in equation (2.43). Note that the angle θ is defined as the angle of inclination of the element measured anticlockwise from the x-axis (at the first node). For this element (Fig. 2.5):

- First node = node 1;
- Second node = node 3;
- Angle $\theta = 30°$;
- Length = 3.464 m.

Note that if the first and second nodes are swapped, the angle θ will be 210°. However, the element stiffness matrix remains the same whether $\theta = 30°$ or 210° is used.

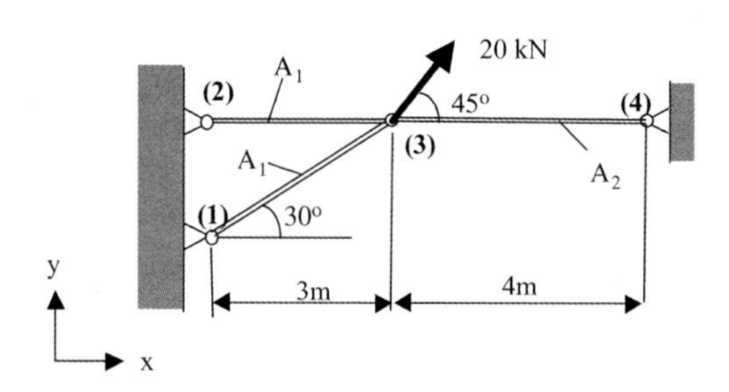

Fig. 2.4 Structural analysis example

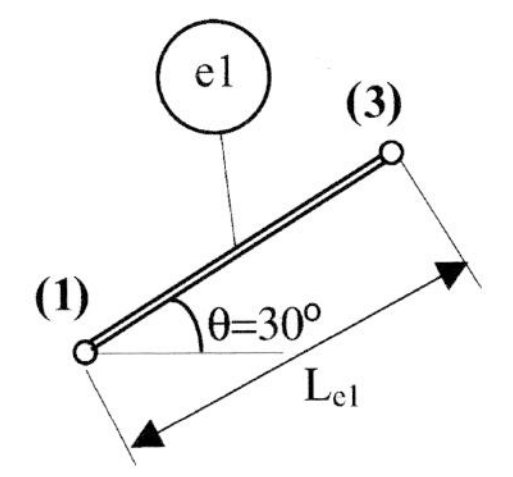

Fig. 2.5 Configuration of element e1

The force–displacement equations for this element can be written as

$$\begin{bmatrix} F_{x1} \\ F_{y1} \\ \hline F_{x3} \\ F_{y3} \end{bmatrix}_{e1} = 10^6 \left[\begin{array}{cc|cc} 11.205 & 6.469 & -11.205 & -6.469 \\ 6.469 & 3.735 & -6.469 & -3.735 \\ \hline -11.205 & -6.469 & 11.205 & 6.469 \\ -6.469 & -3.735 & 6.469 & 3.735 \end{array}\right] \begin{bmatrix} u_{x1} \\ u_{y1} \\ \hline u_{x3} \\ u_{y3} \end{bmatrix}_{e1} \tag{2.51}$$

Putting this expression in the global system of equations results in

$$\begin{bmatrix} F_{x1} \\ F_{y1} \\ \hline F_{x2} \\ F_{y2} \\ \hline F_{x3} \\ F_{y3} \\ \hline F_{x4} \\ F_{y4} \end{bmatrix}_{e1} = 10^6 \left[\begin{array}{cc|cc|cc|cc} 11.205 & 6.469 & X & X & -11.205 & -6.469 & X & X \\ 6.469 & 3.735 & X & X & -6.469 & -3.735 & X & X \\ \hline X & X & X & X & X & X & X & X \\ X & X & X & X & X & X & X & X \\ \hline -11.205 & -6.469 & X & X & 11.205 & 6.469 & X & X \\ -6.469 & -3.735 & X & X & 6.469 & 3.735 & X & X \\ \hline X & X & X & X & X & X & X & X \\ X & X & X & X & X & X & X & X \end{array}\right]$$

$$\times \begin{bmatrix} u_{x1} \\ u_{y1} \\ \hline u_{x2} \\ u_{y2} \\ \hline u_{x3} \\ u_{y3} \\ \hline u_{x4} \\ u_{y4} \end{bmatrix}_{e1} \tag{2.52}$$

where X indicates no contribution to the $[K]$ matrix.

Element stiffness matrix for element e2
Similarly for element e2 (Fig. 2.6):

- First node = node 2;
- Second node = node 3;
- Angle $\theta = 0°$;
- Length = 3 m.

The force–displacement equations for this element can be written as

$$\begin{bmatrix} F_{x2} \\ F_{y2} \\ \hline F_{x3} \\ F_{y3} \end{bmatrix}_{e2} = 10^6 \left[\begin{array}{cc|cc} 17.25 & 0 & -17.25 & 0 \\ 0 & 0 & 0 & 0 \\ \hline -17.25 & 0 & 17.25 & 0 \\ 0 & 0 & 0 & 0 \end{array}\right] \begin{bmatrix} u_{x2} \\ u_{y2} \\ \hline u_{x3} \\ u_{y3} \end{bmatrix}_{e2} \quad (2.53)$$

Putting this expression in the global system of equations results in

$$\begin{bmatrix} F_{x1} \\ F_{y1} \\ \hline F_{x2} \\ F_{y2} \\ \hline F_{x3} \\ F_{y3} \\ \hline F_{x4} \\ F_{y4} \end{bmatrix}_{e2} = 10^6 \left[\begin{array}{cc|cc|cc|cc} X & X & X & X & X & X & X & X \\ X & X & X & X & X & X & X & X \\ \hline X & X & 17.25 & 0 & -17.25 & 0 & X & X \\ X & X & 0 & 0 & 0 & 0 & X & X \\ \hline X & X & -17.25 & 0 & 17.25 & 0 & X & X \\ X & X & 0 & 0 & 0 & 0 & X & X \\ \hline X & X & X & X & X & X & X & X \\ X & X & X & X & X & X & X & X \end{array}\right] \begin{bmatrix} u_{x1} \\ u_{y1} \\ \hline u_{x2} \\ u_{y2} \\ \hline u_{x3} \\ u_{y3} \\ \hline u_{x4} \\ u_{y4} \end{bmatrix}_{e2} \quad (2.54)$$

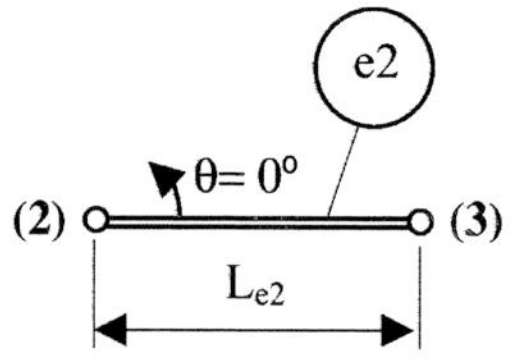

Fig. 2.6 Configuration of element e2

Element stiffness matrix for element e3

For element e3 (Fig. 2.7):

- First node = node 3;
- Second node = node 4;
- Angle $\theta = 0°$;
- Length = 4 m.

The force–displacement equations for this element can be written as

$$\begin{bmatrix} F_{x3} \\ F_{y3} \\ \hline F_{x4} \\ F_{y4} \end{bmatrix}_{e3} = 10^6 \left[\begin{array}{cc|cc} 23.29 & 0 & -23.29 & 0 \\ 0 & 0 & 0 & 0 \\ \hline -23.29 & 0 & 23.29 & 0 \\ 0 & 0 & 0 & 0 \end{array}\right] \begin{bmatrix} u_{x3} \\ u_{y3} \\ \hline u_{x4} \\ u_{y4} \end{bmatrix}_{e3} \tag{2.55}$$

Putting this expression in the global system of equations results in

$$\begin{bmatrix} F_{x1} \\ F_{y1} \\ \hline F_{x2} \\ F_{y2} \\ \hline F_{x3} \\ F_{y3} \\ \hline F_{x4} \\ F_{y4} \end{bmatrix}_{e3} = 10^6 \left[\begin{array}{cc|cc|cc|cc} X & X & X & X & X & X & X & X \\ X & X & X & X & X & X & X & X \\ \hline X & X & X & X & X & X & X & X \\ X & X & X & X & X & X & X & X \\ \hline X & X & X & X & 23.29 & 0 & -23.29 & 0 \\ X & X & X & X & 0 & 0 & 0 & 0 \\ \hline X & X & X & X & -23.29 & 0 & 23.29 & 0 \\ X & X & X & X & 0 & 0 & 0 & 0 \end{array}\right] \begin{bmatrix} u_{x1} \\ u_{y1} \\ \hline u_{x2} \\ u_{y2} \\ \hline u_{x3} \\ u_{y3} \\ \hline u_{x4} \\ u_{y4} \end{bmatrix}_{e3} \tag{2.56}$$

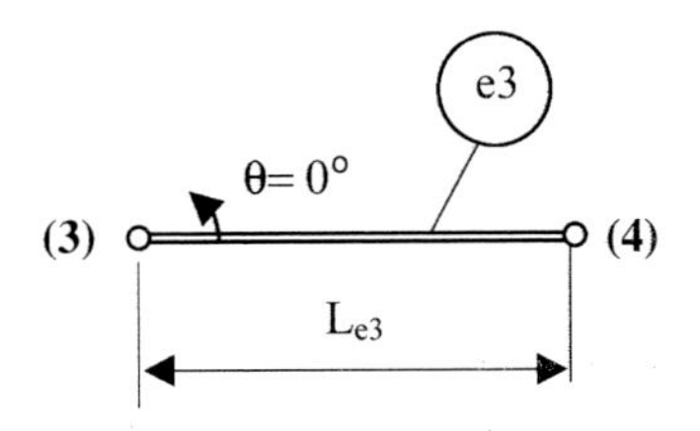

Fig. 2.7 Configuration of element e3

Global assembly of elements

The global force–displacement equations for the whole assembly can be obtained by combining the stiffness matrix contributions of all the individual elements such that the $[K]$ coefficients belonging to common nodes are added together, as shown schematically in Fig. 2.8.

The overall assembled global matrices are

$$
\begin{bmatrix} F_{x1} \\ F_{y1} \\ \hline F_{x2} \\ F_{y2} \\ \hline F_{x3} \\ F_{y3} \\ \hline F_{x4} \\ F_{y4} \end{bmatrix} = 10^6
\times \left[\begin{array}{cc|cc|cc|cc}
11.205 & 6.469 & X & X & -11.205 & -6.469 & X & X \\
6.469 & 3.735 & X & X & -6.469 & -3.735 & X & X \\ \hline
X & X & 17.25 & 0 & -17.25 & 0 & X & X \\
X & X & 0 & 0 & 0 & 0 & X & X \\ \hline
-11.205 & -6.469 & -17.25 & 0 & (11.25+17.25+23.29) & (6.469+0+0) & -23.29 & 0 \\
-6.469 & -3.735 & 0 & 0 & (6.469+0+0) & (3.735+0+0) & 0 & 0 \\ \hline
X & X & X & X & -23.29 & 0 & 23.29 & 0 \\
X & X & X & X & 0 & 0 & 0 & 0
\end{array}\right]
\times \begin{bmatrix} u_{x1} \\ u_{y1} \\ \hline u_{x2} \\ u_{y2} \\ \hline u_{x3} \\ u_{y3} \\ \hline u_{x4} \\ u_{y4} \end{bmatrix}
\tag{2.57}
$$

Note that the $[K]$ coefficients corresponding to node 3, that is, the coefficients multiplying u_{x3} and u_{y3}, now contain three contributions from the three elements connected to node 3.

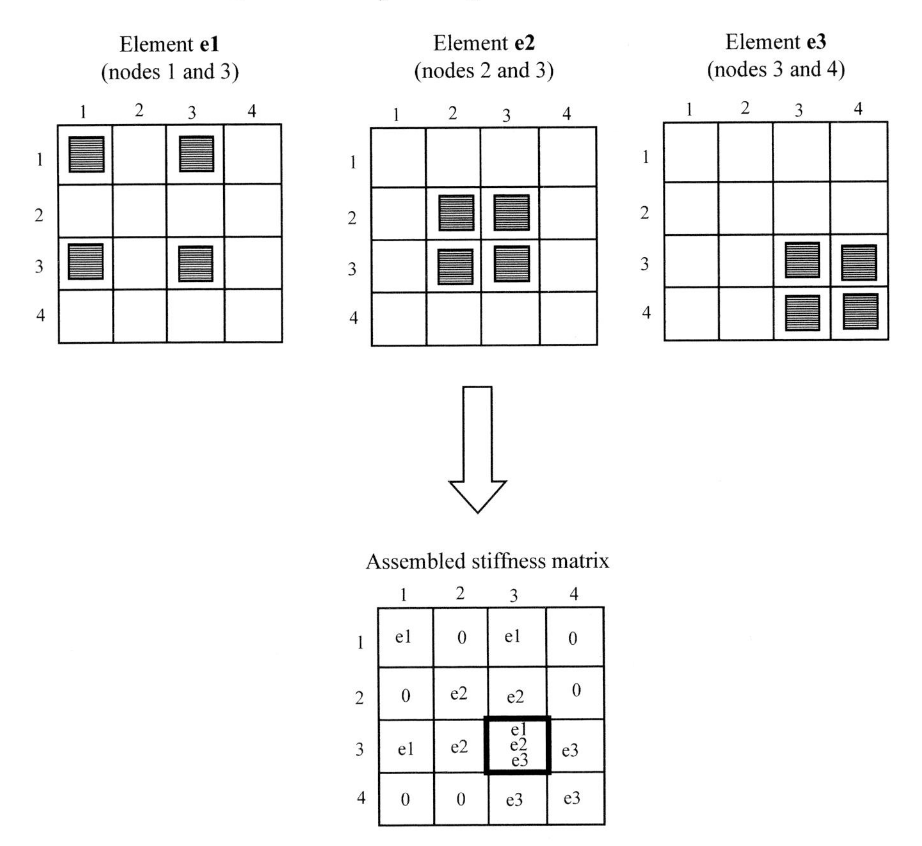

Fig. 2.8 Assembly of the overall stiffness matrix

To obtain a unique solution to the equations, the boundary conditions must be satisfied. In this example, the displacement constraints at points 1, 2, and 4 are

$$
\begin{aligned}
u_{x1} &= u_{y1} = 0 \\
u_{x2} &= u_{y2} = 0 \\
u_{x4} &= u_{y4} = 0
\end{aligned}
\tag{2.58}
$$

and the external force vector has two nonzero components at node 3

$$
\begin{aligned}
F_{x3} &= 20 \cos 45^\circ \times 10^3 \,\mathrm{N} \\
F_{y3} &= 20 \sin 45^\circ \times 10^3 \,\mathrm{N}
\end{aligned}
\tag{2.59}
$$

Therefore the global $[u]$ and $[F]$ vectors can be written as

$$[u]_{\text{global}} = \begin{bmatrix} 0 \\ 0 \\ \hline 0 \\ 0 \\ \hline u_{x3} \\ u_{y3} \\ \hline 0 \\ 0 \end{bmatrix}; \qquad [F]_{\text{global}} = \begin{bmatrix} 0 \\ 0 \\ \hline 0 \\ 0 \\ \hline F_{x3} \\ F_{y3} \\ \hline 0 \\ 0 \end{bmatrix} \tag{2.60}$$

Note that the external force vector does not contain any reaction forces at nodes 1, 2, and 4, because reaction forces are not 'externally' applied forces. The reaction forces will be calculated after solving the overall system of equations.

By multiplying the relevant coefficients in equation (2.57), the eight equations reduce to two equations due to the zeros in the global $[u]$ matrix, as follows.

$$10^3 \begin{bmatrix} 20\cos 45^\circ \\ 20\sin 45^\circ \end{bmatrix} = 10^6 \begin{bmatrix} 51.745 & 6.469 \\ 6.469 & 3.735 \end{bmatrix} \begin{bmatrix} u_{x3} \\ u_{y3} \end{bmatrix} \tag{2.61}$$

from which the displacements of node 3 can be calculated as

$$\begin{aligned} u_{x3} &= -0.256\,\text{mm} \\ u_{y3} &= 4.229\,\text{mm} \end{aligned} \tag{2.62}$$

Note that a negative displacement indicates movement to the left. Once the displacements of the nodal points are computed, other parameters such as element stresses can be calculated using the stress–strain and the strain–displacement relationships, that is,

$$[\sigma_e] = [B][D][u_e] \tag{2.63}$$

Therefore, the stress in element $e1$ can be calculated as

$$[\sigma_{e1}] = \left(\frac{207 \times 10^9}{3.464}\right) \begin{bmatrix} -0.866 & -0.5 & 0.866 & 0.5 \end{bmatrix} \begin{bmatrix} 0 \\ 0 \\ u_{x3} \\ u_{y3} \end{bmatrix}$$

$$= 113.1 \times 10^6\,\text{N/m}^2 \tag{2.64}$$

Similarly, the stresses in elements $e2$ and $e3$ can be calculated as

$$[\sigma_{e2}] = -17.6\,\mathrm{MN/m^2}$$
$$[\sigma_{e3}] = 13.2\,\mathrm{MN/m^2} \tag{2.65}$$

Reaction forces can now be easily calculated by multiplying the element stresses by the relevant cross-sectional area of the element.

2.4 SUMMARY OF KEY POINTS

- The simplest type of element is a one-dimensional (uniaxial) bar element with one degree of freedom (the axial displacement).
- A linear shape function can be used with two-node elements. This means that the displacement is allowed to vary linearly per element, and the stress/strain is constant per element.
- The FE formulation for simple elements can be easily extended to more sophisticated problems, for example, truss elements in two- and three-dimensional problems.
- For each element, a stiffness expression can be derived as

$$[k_e][u_e] = [F_e]$$

- Two alternative approaches can be used to derive the element stiffness matrix: either a direct equilibrium (nonenergy) approach, or an energy approach based on minimizing the total potential energy. The energy approach is more versatile and can be easily adapted for other problems.
- In structural analysis problems, the individual element stiffness matrices can be assembled together in the global stiffness matrix by superimposing the $[K]$ coefficients of the nodes, which are shared between two or more elements.
- The global stiffness matrix $[K]$ is symmetric and sparsely populated.

CHAPTER 3

Continuum Elements (Two-Dimensional, Axisymmetric, Three-Dimensional Elements)

The FE formulation for continuum elements is presented here, including two-dimensional plane strain/plane stress, axisymmetric, and three-dimensional elements. In continuum elements, the degrees of freedom are the displacement components. Displacements are chosen as the independent variables and an approximation for the displacement function within each element is assumed.

3.1 Overview of the FE formulation

The FE formulation used for continuum elements is very similar to that used for the structural analysis discussed earlier. The steps followed in deriving the final system of equations are summarized in Fig. 3.1.

3.2 Two-dimensional and axisymmetric assumptions

3.2.1 Two-dimensional problems

In many applications it is possible to approximate a three-dimensional (3D) problem to a simpler two-dimensional (2D) application in which only the x–y plane is modelled. However, it should be emphasized from the outset that there is no such problem as a truly 2D one; all 2D solutions are approximations of 3D solutions.

Two assumptions about the stress and thickness in the z-direction are made:

(a) *Plane stress* is used to define 'thin' geometries in the z-direction where the stress across the thickness is neglected (i.e. $\sigma_{zz} = 0$), as shown in Fig. 3.2(a).

(b) *Plane strain* is used to define very 'thick' geometries in the z-direction where the strain across the thickness is neglected, but the stress there is nonzero (i.e. $\varepsilon_{zz} = 0$, but $\sigma_{zz} \neq 0$). In other words, the x–y section is remote from the ends where $z = \pm\infty$, as shown in Fig. 3.2(b).

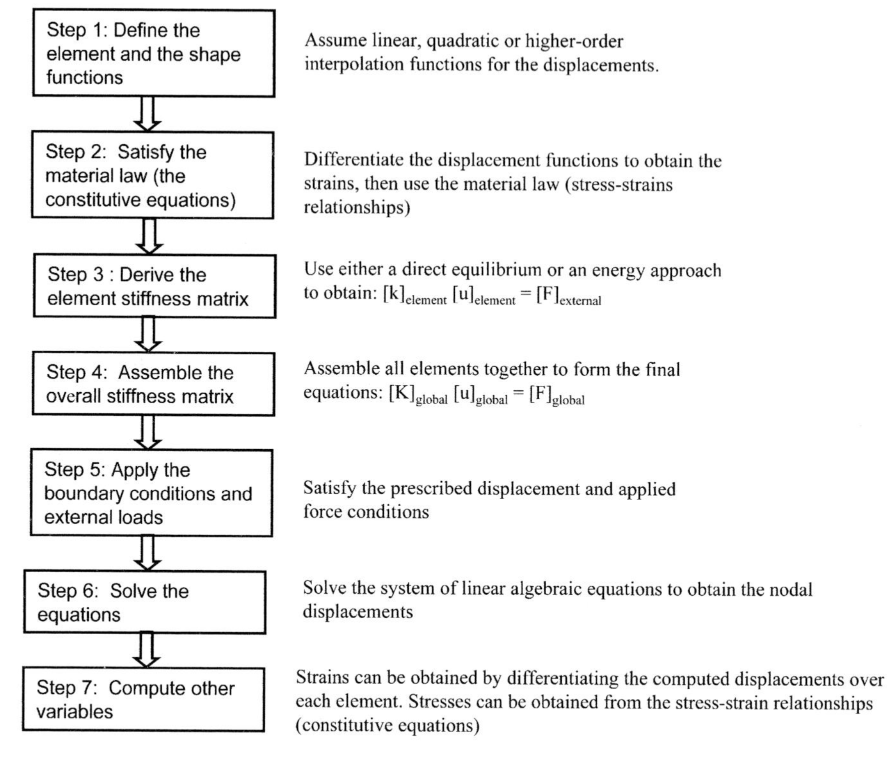

Fig. 3.1 Summary of the derivation of FE formulation

A special case of plane strain arises when thermal strains are present, where the cross-section is prevented from distorting out of plane, but is allowed to have thermal strains in the z-direction. In such cases, usually referred to as 'generalized plane strain', the strain in the z-direction, ε_{zz}, is nonzero.

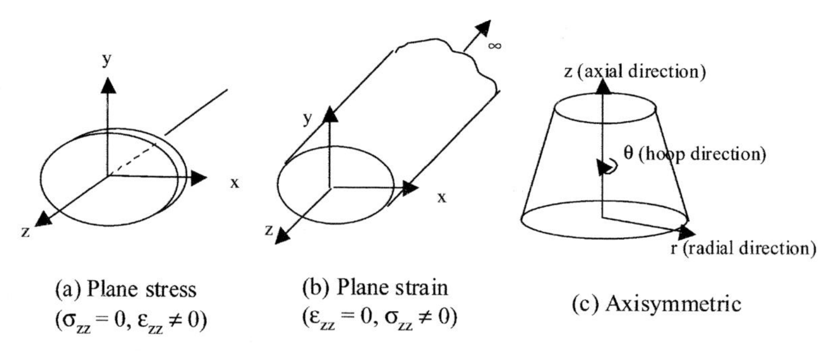

Fig. 3.2 Two-dimensional plane stress, two-dimensional plane strain, and axisymmetric problems

3.2.2 Axisymmetric problems

Axisymmetric problems may also be viewed as two-dimensional, but the x- and y-directions are replaced by the radial (r) and axial (z) direction, as shown in Fig. 3.2(c). Axisymmetric geometries, or bodies of revolution, are formed by rotating a 2D flat plane through 360° about the z-axis. It should be emphasized that for an axisymmetric assumption to be valid, both the geometry and all the variables must be axisymmetric. Therefore, all loads must be ring loads.

3.3 TWO-DIMENSIONAL TRIANGULAR CONTINUUM ELEMENTS

Before deriving the FE formulation, the variables used in 2D formulations should be defined. In 2D continuum problems, these are the displacements $[u]$, strains $[\varepsilon]$, and stresses $[\sigma]$, defined as

$$[u] = \begin{bmatrix} u_x \\ u_y \end{bmatrix}; \qquad [\varepsilon] = \begin{bmatrix} \varepsilon_{xx} \\ \varepsilon_{yy} \\ \varepsilon_{xy} \end{bmatrix}; \qquad [\sigma] = \begin{bmatrix} \sigma_{xx} \\ \sigma_{yy} \\ \sigma_{xy} \end{bmatrix} \tag{3.1}$$

Note that the stress normal to the $x-y$ plane (σ_{zz}) is either zero (plane stress) or nonzero (plane strain).

Step 1: Define the element and the shape functions

The simplest 2D continuum element is a three-node triangle with straight-line sides, as shown in Fig. 3.3(a). This element is called a 'constant strain triangle (CST)' because the strain (and therefore the stress) is constant per

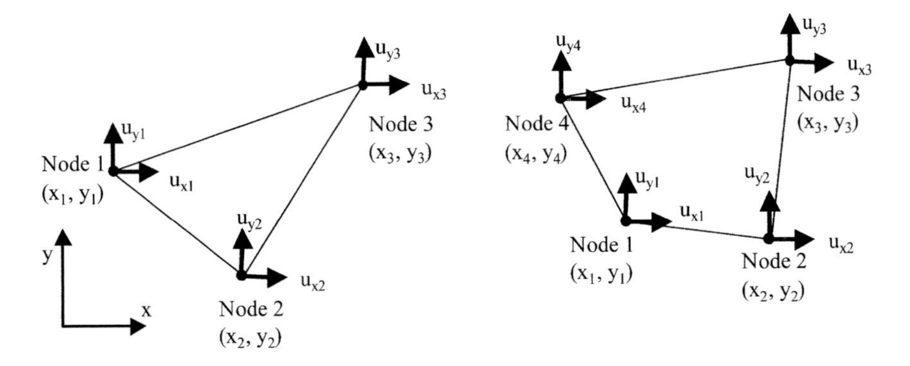

Fig. 3.3 Two-dimensional triangular and quadrilateral elements

element. Alternatively, a quadrilateral four-node straight-sided element can be used, as shown in Fig. 3.3(b).

For simplicity, only the FE formulation for triangular elements is presented here. The same steps and matrix expressions are equally applicable to quadrilateral elements. Assuming a linear variation of displacement over each element, an expression for the displacement in terms of x and y can be written as

$$\begin{aligned} u_x(x, y) &= C_1 + C_2 x + C_3 y \\ u_y(x, y) &= C_4 + C_5 x + C_6 y \end{aligned} \tag{3.2}$$

where C_1 to C_6 are six constants that can be expressed in terms of the coordinates of the nodes, such that

$$\begin{bmatrix} u_{x1} \\ u_{y1} \\ u_{x2} \\ u_{y2} \\ u_{x3} \\ u_{y3} \end{bmatrix} = \begin{bmatrix} 1 & x_1 & y_1 & 0 & 0 & 0 \\ 0 & 0 & 0 & 1 & x_1 & y_1 \\ 1 & x_2 & y_2 & 0 & 0 & 0 \\ 0 & 0 & 0 & 1 & x_2 & y_2 \\ 1 & x_3 & y_3 & 0 & 0 & 0 \\ 0 & 0 & 0 & 1 & x_3 & y_3 \end{bmatrix} \begin{bmatrix} C_1 \\ C_2 \\ C_3 \\ C_4 \\ C_5 \\ C_6 \end{bmatrix} \tag{3.3}$$

These equations can be abbreviated to:

$$[u_e] = [A][C] \tag{3.4}$$

where $[u_e]$ is the displacement vector of the element, $[A]$ is a 'coordinate matrix', and $[C]$ is the vector of the constants C_1 to C_6. Equation (3.4) can be solved to determine the $[C]$ constants via

$$[C] = [A^{-1}][u_e] \tag{3.5}$$

Step 2: Satisfy the material law (constitutive equations)
Using the strain–displacement definitions, the element strain can be determined by differentiating equation (3.2).

$$\begin{aligned} \varepsilon_{xx} &= \frac{\partial u_x}{\partial x} = C_2 \\ \varepsilon_{yy} &= \frac{\partial u_y}{\partial y} = C_6 \\ \varepsilon_{xy} &= \frac{\partial u_x}{\partial y} + \frac{\partial u_y}{\partial x} = C_3 + C_5 \end{aligned} \tag{3.6}$$

The element strain vector $[\varepsilon_e]$ can then be written in matrix form as

$$[\varepsilon] = \begin{bmatrix} \varepsilon_{xx} \\ \varepsilon_{yy} \\ \varepsilon_{xy} \end{bmatrix} = \begin{bmatrix} 0 & 1 & 0 & 0 & 0 & 0 \\ 0 & 0 & 0 & 0 & 0 & 1 \\ 0 & 0 & 1 & 0 & 1 & 0 \end{bmatrix} \begin{bmatrix} C_1 \\ C_2 \\ C_3 \\ C_4 \\ C_5 \\ C_6 \end{bmatrix} \tag{3.7}$$

or, more concisely

$$[\varepsilon] = [X][C] \tag{3.8}$$

Substituting $[C]$ from equation (3.5) into equation (3.8) results in the following expression of element strain in terms of displacement.

$$[\varepsilon] = [X][A^{-1}][u_e] = [B][u_e] \tag{3.9}$$

where $[B]$ is a 'dimension matrix'.

To derive stress–strain relationships, Hooke's law can be used. For 2D problems, plane stress and plane strain problems have to be considered.

Plane stress problems

Here $\sigma_{zz} = 0$, and Hooke's law can be expressed in matrix form as

$$\begin{bmatrix} \varepsilon_{xx} \\ \varepsilon_{yy} \\ \varepsilon_{xy} \end{bmatrix} = \begin{bmatrix} \frac{1}{E} & \frac{-\nu}{E} & 0 \\ \frac{-\nu}{E} & \frac{1}{E} & 0 \\ 0 & 0 & \frac{1+\nu}{E} \end{bmatrix} \begin{bmatrix} \sigma_{xx} \\ \sigma_{yy} \\ \sigma_{xy} \end{bmatrix} \tag{3.10}$$

Stresses can be written in terms of strains by algebraic manipulation of the above equation, as

$$[\sigma] = [D][\varepsilon] \tag{3.11}$$

where $[D]$ is a 'material property matrix' defined as

$$[D] = \left(\frac{E}{1-\nu^2}\right) \begin{bmatrix} 1 & \nu & 0 \\ \nu & 1 & 0 \\ 0 & 0 & 1-\nu \end{bmatrix} (\textit{plane stress}) \tag{3.12}$$

Note that matrix $[D]$ is symmetric, that is, $[D]^T = [D]$.

Therefore, using equation (3.9), the stress vector $[\sigma]$ can be written in terms of the displacement vector $[u_e]$ as

$$[\sigma] = [D][B][u_e] \tag{3.13}$$

Plane strain problems
For plane strain problems, $\varepsilon_{zz} = 0$, and the σ_{zz} stress is nonzero, determined as

$$\sigma_{zz} = \nu(\sigma_{xx} + \sigma_{yy}) \tag{3.14}$$

Eliminating σ_{zz} from the Hooke's law expressions gives the strain expression

$$\begin{bmatrix} \varepsilon_{xx} \\ \varepsilon_{yy} \\ \varepsilon_{xy} \end{bmatrix} = \begin{bmatrix} \dfrac{1-\nu^2}{E} & \dfrac{-\nu(1+\nu)}{E} & 0 \\ \dfrac{-\nu(1+\nu)}{E} & \dfrac{1-\nu^2}{E} & 0 \\ 0 & 0 & \dfrac{1+\nu}{E} \end{bmatrix} \begin{bmatrix} \sigma_{xx} \\ \sigma_{yy} \\ \sigma_{xy} \end{bmatrix} \tag{3.15}$$

Stresses can be written in terms of strains by algebraic manipulation of the above equation, as

$$[\sigma] = [D][\varepsilon] \tag{3.16}$$

where $[D]$ is a 'material property matrix' defined as

$$[D] = \frac{E}{(1-2\nu)(1+\nu)} \begin{bmatrix} 1-\nu & \nu & 0 \\ \nu & 1-\nu & 0 \\ 0 & 0 & 1-2\nu \end{bmatrix} (\textit{plane strain}) \tag{3.17}$$

Special case of an incompressible material
It is worth examining the special case of an incompressible material when Poisson's ratio is equal to or close to 0.5. For plane stress problems, this poses no problems. However, for plane strain problem, $\nu = 0.5$ results in the denominator of $[D]$ being zero, that is, the matrix $[D]$ becomes 'singular'. This clearly demonstrates that the FE formulation for 2D plane strain incompressible or nearly incompressible materials is not valid when the displacements are used as the independent variables. For this reason, 'hybrid elements' are used for such materials, where extra variables (pressures) are used as the independent variables. Hybrid elements use more internal variables over each element, and consequently consume more computation time.

Step 3: Derive the element stiffness matrix
The stiffness matrix can be derived using an energy (variational) formulation, which is considered more robust than the direct equilibrium approach. The advantage of using the energy formulation is that it can be easily adapted for any type of element (e.g. quadratic, three-dimensional).

The total potential energy (TPE) of the element can be expressed as

$$TPE = \int_v \frac{1}{2}[\sigma]^T[\varepsilon]\,\mathrm{d}v - [u_e]^T[F_e] \tag{3.18}$$

where v is the volume of the element. Substituting for the stress and strain from equations (3.16) and (3.9), respectively, results in

$$\begin{aligned} TPE &= \int_v \frac{1}{2}([D][B][u_e])^T([B][u_e])\,\mathrm{d}v - [u_e]^T[F_e] \\ &= \int_v \frac{1}{2}[u_e]^T\,[B]^T\,[D][B][u_e]\,\mathrm{d}v - [u_e]^T[F_e] \end{aligned} \tag{3.19}$$

Using the principle of minimum total potential energy, the differential of TPE with respect to the displacement $[u]$ must be zero, that is,

$$\frac{\partial(TPE)}{\partial[u_e]} = 0 = \int_v [B]^T\,[D][B][u_e]\,\mathrm{d}v - [F_e] \tag{3.20}$$

which can be expressed in terms of the element stiffness matrix $[k_e]$, as

$$[k_e][u_e] = [F_e] \tag{3.21}$$

where

$$[k_e] = \int_v [B]^T[D][B]\,\mathrm{d}v \tag{3.22}$$

Step 4: Assemble the overall stiffness matrix
To form the overall stiffness matrix, all the energy contributions from the individual elements are added together as

$$\left(\frac{\partial(TPE)}{\partial u}\right)_{\text{overall}} = \sum_{\text{All elements}} \left(\frac{\partial(TPE)}{\partial u}\right)_{\text{element}} = 0 \tag{3.23}$$

This assumes that element interfaces do not contribute to the overall energy of the structure. The final global system of equations is, therefore, arrived at by

$$[K]_{\text{global}}\,[u]_{\text{global}} = [F]_{\text{global}} \tag{3.24}$$

where $[K]$ is the global stiffness matrix. For a given 2D problem with N nodes, with each node having two degrees of freedom (here u_x and u_y), the overall $[K]$ matrix is of size $(2N \times 2N)$, while the global $[u]$ and $[F]$ vectors are of size $(2N \times 1)$.

Alternatively, a direct equilibrium approach may be used to derive an overall stiffness matrix by equating the summation of the internal forces on the nodes to the externally applied forces. For each node, the summation of the internal forces over all the elements that contain the node, must be equal to the external forces applied at that node (if any).

It should be emphasized that the summation process is performed only over elements that share a particular node. Therefore, the overall $[K]$ is sparsely populated, that is, with very few nonzero coefficients. For example, for a $[K]$ matrix of size 1000×1000 (i.e. 500 nodes), assuming no more than, say, nine elements can be connected to a given node, then the maximum number of nonzero coefficients in any given row of 1000 coefficients is 18.

Step 5: Apply the boundary conditions and external loads

To obtain a unique solution for the displacement at every node, some displacement constraints and loads must be specified. The simplest nodal conditions are either prescribed displacements (constraints) or prescribed loads, as described below.

Prescribed loads

This is easily accommodated in the overall equations by specifying the magnitudes of the forces in the appropriate locations of the overall external force vector $[F]$. For example, if node n has a prescribed force P at an angle θ to the x-axis, the force vector for this node (rows $2n - 1$ and $2n$) becomes

$$\begin{bmatrix} F_x \\ F_y \end{bmatrix}_n = \begin{bmatrix} P\cos\theta \\ P\sin\theta \end{bmatrix} \tag{3.25}$$

If no load is specified for a given node, then a zero value should appear in that node's location in the $[F]$ vector.

Prescribed displacements (constraints)

If a displacement value is prescribed for a given node, then there is no need to calculate the displacement value at that node (the equation is effectively eliminated from the solver). However, it is much more convenient to incorporate this condition in the overall system of equations without disturbing the solver. This is easily done by making the diagonal coefficient that multiplies the prescribed displacement equal to one, and the rest of the coefficients in the relevant row equal to zero. This ensures that a standard solver can be used directly without modification.

For example, if node n has a prescribed nonzero displacement of δ at an angle θ to the x-axis, then implementing the above arrangement will result in the following two equations (rows $(2n-1)$ and $2n$).

$$\begin{bmatrix} u_x \\ u_y \end{bmatrix}_n = \begin{bmatrix} \delta \cos\theta \\ \delta \sin\theta \end{bmatrix} \tag{3.26}$$

Step 6: Solve the equations

The final system of simultaneous algebraic equations can be solved by any standard solver such as the Gaussian elimination technique. Because the number of unknowns is equal to the number of equations, a unique solution is obtained. However, because the $[K]$ matrix is sparsely populated, storing the full $[K]$ matrix (with all the zero coefficients) is very wasteful of computer storage space.

Special techniques can be used to condense the $[K]$ matrix such that only the nonzero coefficients are stored, resulting in a substantial reduction in the storage requirements. These techniques usually employ routines to minimize the size of the 'bandwidth' (the maximum number of nonzero elements in a given row). The 'frontal' method is a special Gaussian elimination procedure used to solve a set of nodal equations while minimizing the storage requirements.

Step 7: Compute other variables

The strains can be obtained by differentiating the displacements over each element. The stresses can then be derived from the strains using the material law (the constitutive equations).

3.4 AXISYMMETRIC (RING) CONTINUUM ELEMENTS

As discussed earlier in Section 3.2, an axisymmetric problem can be represented by a 2D r–z plane rotated through 360° about the z-axis. The variables in axisymmetric problems are

$$[u] = \begin{bmatrix} u_r \\ u_z \end{bmatrix}; \qquad [\varepsilon] = \begin{bmatrix} \varepsilon_{rr} \\ \varepsilon_{zz} \\ \varepsilon_{rz} \end{bmatrix}; \qquad [\sigma] = \begin{bmatrix} \sigma_{rr} \\ \sigma_{zz} \\ \sigma_{rz} \end{bmatrix} \tag{3.27}$$

Note that the hoop stress $\sigma_{\theta\theta}$ is perpendicular to the r–z plane. The shear stresses $\sigma_{r\theta}$ and $\sigma_{z\theta}$ are zero. The strain–displacement definitions are the same as those used in 2D problems, except that x and y are replaced by r and z.

$$[\varepsilon] = \begin{bmatrix} \varepsilon_{rr} \\ \varepsilon_{zz} \\ \varepsilon_{rz} \end{bmatrix} \begin{bmatrix} \dfrac{\partial u_r}{\partial r} \\ \dfrac{\partial u_z}{\partial z} \\ \dfrac{\partial u_r}{\partial z} + \dfrac{\partial u_z}{\partial r} \end{bmatrix} \tag{3.28}$$

The hoop strain is defined as

$$\varepsilon_{\theta\theta} = \frac{u_r}{r} \tag{3.29}$$

The FE formulation for axisymmetric problems follows the same steps as those for 2D problems with triangular or quadrilateral elements. Axisymmetric elements, however, should be regarded as ring elements, as shown in Fig. 3.4.

The material property matrix $[D]$ used in the axisymmetric FE formulation is the same as the 2D plane strain matrix in equation (3.17).

It should be emphasized that axisymmetric problems must have an axisymmetric geometry and axisymmetric boundary conditions and loads, that is, ring loads. If an axisymmetric geometry has nonaxisymmetric (arbitrary) boundary conditions, then it should be treated as a 3D continuum problem. An example is shown in Fig. 3.5 where a chimney or a cooling tower is subjected to wind load at its side. Some FE codes use special

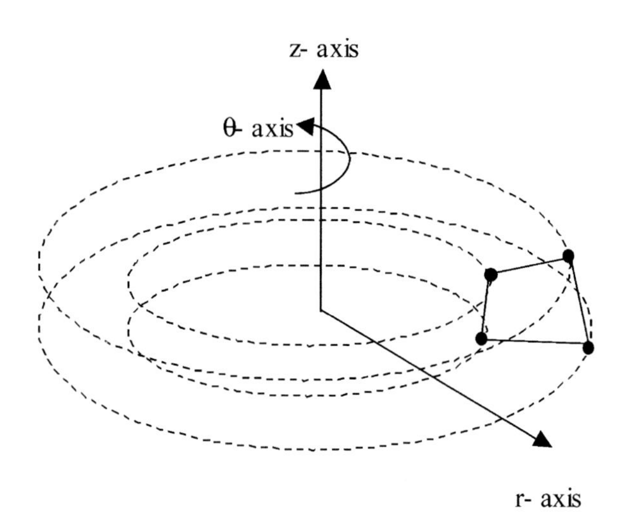

Fig. 3.4 Axisymmetric (ring) continuum element

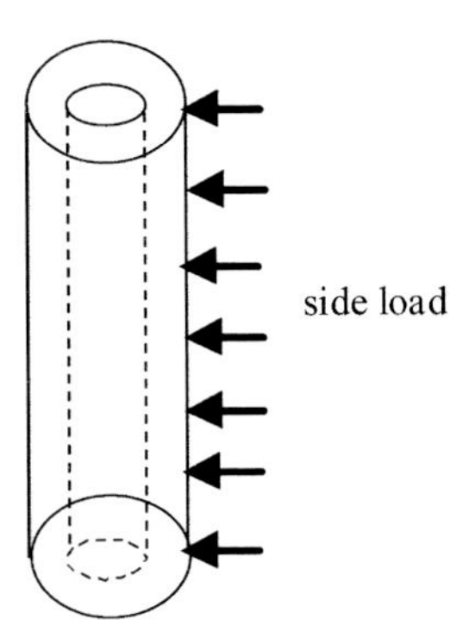

Fig. 3.5 An axisymmetric geometry with nonaxisymmetric boundary conditions

axisymmetric elements that allow arbitrary boundary conditions, in which the nonaxisymmetric loads and displacements are prescribed as a Fourier series. These Fourier elements, however, are not widely used, since they require a good knowledge of Fourier series, and are only efficient for loads that do not vary sharply around the circumference of the axisymmetric geometry.

3.5 THREE-DIMENSIONAL CONTINUUM ELEMENTS

Three-dimensional continuum elements can be used to model all practical engineering problems because no dimensional approximation is assumed in representing the geometry and loads. However, the computational cost associated with 3D problems is considerably higher than that of 2D problems. For example, for a 2D problem with 100 nodes and 2 degrees of freedom per node, 200 equations will be generated. For a similar spread of nodes in a 3D problem (i.e. $10 \times 10 \times 10$ nodes) 3000 equations will be generated.

Typical 3D linear continuum elements are shown in Fig. 3.6. The simplest 3D element is a four-node constant-strain tetrahedron with a linear variation of displacement as follows.

$$\begin{aligned} u_x(x, y, z) &= C_1 + C_2x + C_3y + C_4z \\ u_y(x, y, z) &= C_5 + C_6x + C_7y + C_8z \\ u_z(x, y, z) &= C_9 + C_{10}x + C_{11}y + C_{12}z \end{aligned} \tag{3.30}$$

Derivation of the 3D FE formulation follows the same steps as those used previously for other elements, with the same matrix notation, except that the

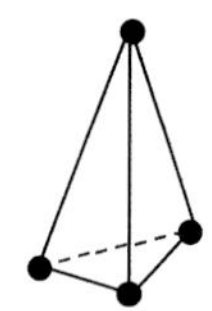

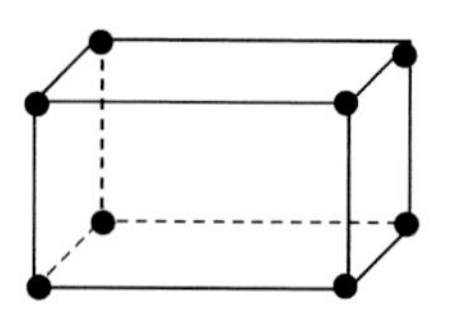

(a) 4-node tetrahedron element (b) 8-node hexahedron element

Fig. 3.6 Three-dimensional (linear) continuum elements

3D matrices are larger than the corresponding 2D matrices. The material law for 3D problems can be rearranged such that the stresses are on the left-hand side, resulting in Hooke's law for 3D stresses as

$$\begin{bmatrix} \sigma_{xx} \\ \sigma_{yy} \\ \sigma_{zz} \\ \sigma_{xy} \\ \sigma_{xz} \\ \sigma_{yz} \end{bmatrix} = \frac{E}{(1-2v)(1+v)}$$

$$\times \begin{bmatrix} 1-v & v & v & 0 & 0 & 0 \\ v & 1-v & v & 0 & 0 & 0 \\ v & v & 1-v & 0 & 0 & 0 \\ 0 & 0 & 0 & \dfrac{(1-2v)}{2} & 0 & 0 \\ 0 & 0 & 0 & 0 & \dfrac{(1-2v)}{2} & 0 \\ 0 & 0 & 0 & 0 & 0 & \dfrac{(1-2v)}{2} \end{bmatrix} \begin{bmatrix} \varepsilon_{xx} \\ \varepsilon_{yy} \\ \varepsilon_{zz} \\ \varepsilon_{xy} \\ \varepsilon_{xz} \\ \varepsilon_{yz} \end{bmatrix}$$

Note that, as in 2D plane strain problems, incompressible materials with Poisson's ratio close to or equal to 0.5 will cause the $[D]$ matrix to become singular.

3.6 SUMMARY OF KEY POINTS

- Triangular three-node elements with a linear variation of displacements (linear shape functions) are the simplest type of 2D continuum elements.

- In continuum elements, the displacement vectors are the independent variables (degrees of freedom). In 2D continuum elements, each node has two degrees of freedom, the displacement vectors u_x and u_y in the x- and y-directions, respectively, whereas in 3D continuum elements, a third degree of freedom, u_z, is used.
- The derivation of the FE formulation for 2D, axisymmetric, and 3D continuum elements follows the same steps.
- In practical engineering problems, the overall stiffness matrix (the solution matrix) is large in size, but is sparsely populated.

CHAPTER 4

Energy and Variational Approaches

4.1 INTRODUCTION

Most FE formulations are derived using a generalized energy or 'variational' formulation rather than a direct equilibrium approach based on the equilibrium of forces. There are many advantages in using variational approaches in preference to direct equilibrium approaches, including the following:

- Variational formulations are more general and cover a wider range of problems, including nonstructural problems such as heat transfer.
- Variational methods can be easily adapted to incorporate different types of elements (e.g. truss, beam, shell, triangular, etc.) and dimensionalities (e.g. 2D, 3D, axisymmetric).
- Quantities used in the variational formulations are scalar quantities (such as energy, potentials, etc.), which are easier to handle than vector quantities (such as displacements and forces).
- Variational methods provide a better understanding of the accuracy and degree of approximation in the final solutions.

A variational approach is usually used to solve the governing partial differential equations (PDE) of structural mechanics or any physical problem, by finding the conditions that make some quantity (usually energy) stationary (either maximum or minimum). This quantity is often referred to as the 'functional', which usually involves the integration of the unknowns of the problem.

In elasticity problems, the functional is the total potential energy (*TPE*) of the structure, and the terms 'variational methods' and 'energy methods' are often used interchangeably. In problems not involving elasticity, such as heat conduction problems, the term 'variational' is more appropriate because it is often difficult to formulate the problem in terms of energy.

The following sections describe the different forms of the variational formulation as used in FE formulations. These include the Ritz and the Galerkin weighted residual techniques.

4.2 RITZ METHOD (VARIATIONAL APPROACH)

Also known as the Rayleigh–Ritz method, this method involves seeking approximate solutions to the governing partial differential equations (PDE) by using an approximate (trial) series solution. The Ritz approach in the FE formulation usually involves the use of a polynomial series solution for the unknown variables of the problem (usually displacements). Each term in the series is associated with an arbitrary constant. These constants, which are finite in number, become the unknowns of the problem, and can be obtained using the appropriate variational principle. The Ritz method can produce accurate results provided certain convergence conditions are met.

4.2.1 Solution steps in the Ritz method

The main steps involved in the Ritz method are outlined below.

Step 1

Assume a trial function that is capable of representing the variables of the problem (usually displacements). The trial solution must satisfy the essential (displacement) boundary conditions of the problem. However, the natural (force) boundary conditions are implicitly satisfied in the *TPE* derivation. For example, the displacement function may be written as

$$u(x) = C_1 + C_2 x + C_3 x^2 \tag{4.1}$$

where C_1, C_2, and C_3 are constants.

Step 2

Derive an integral equation for the total potential energy (*TPE*) in terms of the variables (i.e. displacements) of the problem, that is,

$$TPE = \int f(u)\,\mathrm{d}x \tag{4.2}$$

Step 3

Minimize the *TPE* with respect to the parameters of the trial solution (i.e. with respect to all the C constants), that is,

$$\frac{\partial(TPE)}{\partial C_1} = \frac{\partial(TPE)}{\partial C_2} = \frac{\partial(TPE)}{\partial C_3} = 0 \tag{4.3}$$

Step 4

Solve the resulting equations to obtain all the parameters used in the trial solution, that is, the C_i constants.

4.2.2 Essential and natural boundary conditions

To ensure a unique solution for a given problem, all the boundary conditions must be satisfied, and the strains and displacements must be compatible between adjoining elements. There are two types of boundary conditions in structural mechanics problems:

(a) essential boundary conditions, also called 'geometric' boundary conditions, which refer to displacement and rotation (slope) conditions;
(b) natural boundary conditions, which refer to force and moment conditions.

Using the minimum *TPE* principle is equivalent to satisfying the equations of equilibrium, because the material properties (constitutive equations) are used to obtain stress–strain relationships, and the compatibility and boundary conditions are satisfied.

4.2.3 Convergence criteria

Convergence of the Ritz method means that as the number of terms in the series solution increases, the Ritz trial solution should converge to the exact solution. This will occur provided the following convergence criteria are met:

(a) The set of trial functions must be general enough to cover the exact solution of the problem, that is, they must be general enough to cover the actual physical problem. For example, if the solution is symmetrical about a given axis, then the trial solution must also be symmetrical about the same axis.
(b) Each individual trial function must give rise to finite terms in the *TPE* (or functional) to be minimized.
(c) Each trial function must be physically acceptable; that is, it must satisfy compatibility requirements.
(d) Each trial function must satisfy the essential boundary conditions (i.e. displacement conditions) either individually or as a whole. The trial function need not satisfy the natural boundary conditions (i.e. force conditions) because they are approximately or implicitly satisfied by the minimization of the *TPE* or the functional.

4.3 RITZ EXAMPLE: CANTILEVER BEAM EXAMPLE

This example is chosen to demonstrate the application of the Ritz method to a practical engineering problem – the deflection of a cantilever beam under a uniformly distributed load. To illustrate the degree of approximation involved in the Ritz method, more than one trial solution is used, and the degree of accuracy associated with each trial solution is established by comparing it with the exact solution.

Consider a cantilever beam of length L under a uniformly distributed load w_o, as shown in Fig. 4.1. The objective of the analysis is to derive an

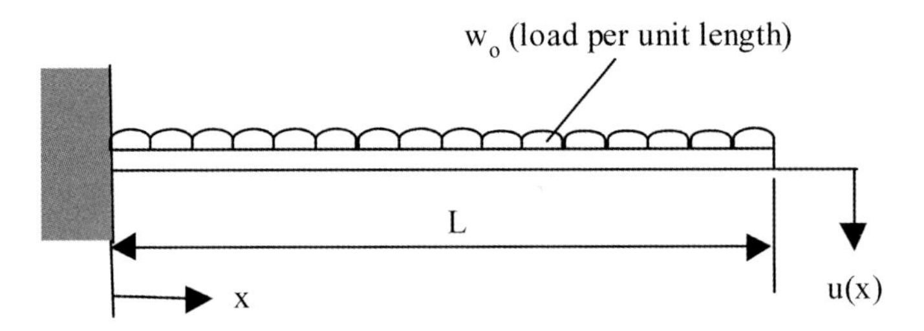

Fig. 4.1 Cantilever beam example

expression for the downward deflection of the beam, $u(x)$, using the Ritz method, that is, the principle of minimum *TPE*.

The first step in the Ritz solution is to assume a trial solution that satisfies the essential (i.e. displacement) boundary conditions.

Essential boundary conditions

In this problem, there are only two essential boundary conditions: zero deflection and zero slope at the built-in end:

(a) At $x = 0$, $u = 0$;
(b) At $x = 0$, $\mathrm{d}u/\mathrm{d}x = 0$.

Natural boundary conditions

It is instructive also to examine whether the Ritz trial solutions satisfy the natural boundary conditions. In the cantilever problem, the natural boundary condition is the zero bending moment condition at the free end of the beam, which can be written as

$$M = EI\left(\frac{\mathrm{d}^2 u}{\mathrm{d}x^2}\right) = 0 \quad \text{at } x = L \tag{4.4}$$

Total potential energy expression

For a cantilever beam under bending, the *TPE* can be derived from classical beam bending theory (**8**) as

$$TPE = \int_0^L \left[\frac{1}{2} EI\left(\frac{\mathrm{d}^2 u}{\mathrm{d}x^2}\right)^2 - w_o u\right] \mathrm{d}x \tag{4.5}$$

where E is Young's modulus and I is the second moment of area.

One-term trial solution

The simplest one-term trial solution that satisfies the displacement boundary conditions can be written as

$$u(x) = C_1\left(\frac{x}{L}\right)^2 \tag{4.6}$$

Substituting this one-term expression in the *TPE* expression gives

$$TPE = \int_0^L \left[\frac{1}{2}EI\left(\frac{2C_1}{L^2}\right)^2 - w_o C_1 \frac{x^2}{L^2}\right] \mathrm{d}x \tag{4.7}$$

which can be integrated to obtain

$$TPE = 2EI\frac{C_1^2}{L^3} - \frac{w_o L}{3}C_1 \tag{4.8}$$

Applying the principle of minimum *TPE*, that is, minimizing the *TPE*, results in

$$\frac{\partial(TPE)}{\partial C_1} = 0 = 4EI\frac{C_1}{L^3} - \frac{w_o L}{3} \tag{4.9}$$

Hence, C_1 can be obtained from the above expression as

$$C_1 = \frac{w_o L^4}{12EI} \tag{4.10}$$

By substituting C_1 into the displacement expression, the following displacement solution is obtained.

$$u = \frac{w_o L^2}{12EI}x^2 \tag{4.11}$$

The value of the *TPE* can be calculated by substituting C_1 into the *TPE* expression in equation (4.8) such that

$$TPE = \frac{2EI}{L^3}\left(\frac{w_o L^4}{12EI}\right)^2 - \frac{w_o L}{3}\left(\frac{w_o L^4}{12EI}\right) = -0.0139\frac{w_o^2 L^5}{EI} \tag{4.12}$$

Note that the value of *TPE* is negative. To check whether this trial solution satisfies the natural (moment) boundary condition at $x = L$, the displacement expression is differentiated twice with respect to x and then substituted in equation (4.4) to give

$$M = EI\left(\frac{d^2u}{dx^2}\right) = EI\left(\frac{2C_1}{L^2}\right) = 0.167w_oL^2 \tag{4.13}$$

which is nonzero; that is, the natural boundary condition is not satisfied by the one-term trial solution.

Two-term trial solution

A more sophisticated two-term trial solution that also satisfies the displacement boundary conditions can be written as

$$u(x) = C_1\left(\frac{x}{L}\right)^2 + C_2\left(\frac{x}{L}\right)^3 \tag{4.14}$$

By repeating the Ritz solution steps for the two-term trial solution, the following values for C_1 and C_2 are obtained.

$$C_1 = \frac{5w_oL^4}{24EI}; \qquad C_2 = \frac{-w_oL^4}{12EI} \tag{4.15}$$

Therefore, the two-term Ritz solution for the displacement is

$$u = \frac{w_oL^2}{24EI}\left(5x^2 - \frac{2x^3}{L}\right) \tag{4.16}$$

and

$$TPE = -0.0243\frac{w_o^2L^5}{EI} \tag{4.17}$$

To check whether this trial solution satisfies the natural (moment) boundary condition at $x = L$, the displacement expression is differentiated twice with respect to x and then substituted in equation (4.4) to give

$$M = -0.0833w_oL^2 \tag{4.18}$$

which is nonzero. Therefore, the two-term solution does not satisfy the natural boundary condition.

Three-term trial solution

A three-term trial solution that also satisfies the displacement boundary conditions can be written as

$$u(x) = C_1\left(\frac{x}{L}\right)^2 + C_2\left(\frac{x}{L}\right)^3 + C_3\left(\frac{x}{L}\right)^4 \tag{4.19}$$

By repeating the Ritz solution steps for the three-term trial solution, the constants can be obtained as

$$C_1 = \frac{w_o L^4}{4EI}; \qquad C_2 = \frac{-w_o L^4}{6EI}; \qquad C_3 = \frac{w_o L^4}{24EI} \tag{4.20}$$

Substituting the C_1, C_2, and C_3 constants in the three-term Ritz solution for displacement gives

$$u = \frac{w_o L^2}{24EI}\left(6x^2 - \frac{4x^3}{L} + \frac{x^4}{L^2}\right) \tag{4.21}$$

The value of *TPE* can be calculated by substituting C_1, C_2, and C_3 into the *TPE* expression, such that

$$TPE = -0.025\frac{w_o^2 L^5}{EI} \tag{4.22}$$

To check whether this trial solution satisfies the natural (moment) boundary condition at $x = L$, the displacement expression is differentiated twice with respect to x and then substituted in equation (4.4) to give

$$M = 0 \tag{4.23}$$

Therefore, the three-term trial solution exactly satisfies the natural boundary condition.

Comparison with the exact solution

The exact analytical solution for the deflection of a cantilever beam under a uniformly distributed load, can be derived from classical beam bending theory (**9**) as

$$u = \frac{w_o L^2}{24EI}\left[\left(1 - \frac{x}{L}\right)^4 + \frac{4x}{L} - 1\right] \tag{4.24}$$

It is worth noting that the above exact solution is different from all the trial solutions used above. It is therefore useful to compare the displacement solutions obtained by the Ritz method to the exact solution, as shown in Table 4.1.

Some useful observations, which are also applicable to other structural mechanics problems, can be made as follows:

- The higher order solutions in which more terms are used produce more accurate solutions.
- Using a higher order solution results in a smaller value of *TPE* (i.e. a better minimization of *TPE*).

Table 4.1 Comparison of Ritz and exact solutions for the cantilever problem

	One-term trial solution	*Two-term trial solution*	*Three-term trial solution*	*Exact solution*
Displacement solution	Of order (x^2)	Of order (x^3)	Of order (x^4)	Of order (x^4)
Displacement BC (Essential BC)	Exactly satisfied	Exactly satisfied	Exactly satisfied	Exactly satisfied
Moment BC (Natural BC)	Approximately satisfied	Approximately satisfied	Exactly satisfied	Exactly satisfied
Deflection at $x = 0$	0	0	0	0
Deflection ($w_o^2 L^4/EI$) at $x/L = 0.25$	0.00521	0.0117	0.0132	0.0132
Deflection ($w_o^2 L^4/EI$) at $x/L = 0.5$	0.0208	0.0417	0.0443	0.0443
Deflection ($w_o^2 L^4/EI$) at $x/L = 1$	0.0833	0.125	0.125	0.125
Value of *TPE* ($w_o^2 L^5/EI$)	−0.0139	−0.0243	−0.025	–

BC, boundary condition; *TPE*, total potential energy.

- Although the displacement boundary conditions are satisfied in the trial solution, the natural boundary conditions (i.e. force and moment conditions) are not necessarily exactly satisfied.
- A trial solution will be more accurate if it satisfies *both* the essential and natural boundary conditions. However, in practical engineering problems it is often very difficult to find such functions, and it is therefore more effective to use more terms in the trial solution that satisfy only the essential boundary conditions.

4.4 GALERKIN METHOD (WEIGHTED RESIDUALS)

The main disadvantage of the Ritz method is that it does not operate *directly* on the governing partial differential equations (PDE) of the physical problem. Instead, it formulates the problem in terms of a variational or energy principle. In some physical problems, however, the use of a variational principle may not be possible or convenient, and other approximating methods may be more suitable than the Ritz method.

In order to apply the FE method to problems other than structural problems, it becomes necessary to use a more generalized procedure for deriving the solutions for the displacements. These methods are often called the 'weighted residual' methods, and derive the solutions directly from the partial differential equations (PDE) of the problem without the need for minimizing a functional or the *TPE*.

The PDE for a given problem is usually written as a differential equation with zero on the right-hand side, that is

$$PDE = 0 \tag{4.25}$$

where *PDE* represents the left-hand side of a partial differential equation.

The PDE can be used to provide a measure of the error or 'residual' of the approximation by substituting the trial solution and checking that the right-hand side of the equation is as close to zero as possible. Therefore, the residual or error can be expressed as

$$Residual(x) = PDE(trial\ solution) \tag{4.26}$$

Therefore, the best solution is that which minimizes the $residual(x)$ function. Assuming that the trial solution for the displacement contains M terms with M constants (i.e. the C_i unknowns), it is necessary to create M equations in order to solve for the M unknowns.

There are a number of alternative ways of minimizing the residual and obtaining M simultaneous equations corresponding to the C_i parameters:

(a) The residual is made equal to zero at M arbitrary points (called the 'Collocation' method).

(b) The 'mean square error' is minimized with respect to all the C_i coefficients, as

$$\int_0^L [residual(x)]^2 \, \mathrm{d}x = 0 \tag{4.27}$$

(c) Instead of making the residual equal to zero, it is multiplied by appropriate functions called 'weight functions' to satisfy M conditions of the form

$$\int_0^L residual(x)\, \phi(x) \, \mathrm{d}x = 0 \tag{4.28}$$

where the ϕ_i functions are weighting functions. This approach is called the 'weighted residual method'.

The Galerkin weighted residual method is widely used in association with the FE method. In the weighted residual integral expression, the weighting functions, $\phi(x)$, are the same functions as the functions used in the Ritz trial solution.

The main steps involved in the Galerkin method are outlined below.

Step 1

As in the Ritz method, assume a trial function that is capable of representing the variables of the problem (usually displacement). The trial solution must satisfy the boundary conditions of the problem.

For example, the displacement function may be written as

$$u(x) = C_1 + C_2 x + C_3 x^2 \tag{4.29}$$

where C_1, C_2, and C_3 are constants.

Step 2

Derive the governing partial differential equations (PDE), for example, in structural problems, by considering the equilibrium of a small element of the structure, such that

$$PDE = f(u) = 0 \tag{4.30}$$

Step 3

Use a weighted residual integral expression in which the weighting functions, $\phi(x)$, used are equal to the trial functions, such that

$$\int_0^L (PDE)(trial\ solution)\,\mathrm{d}x = 0 \tag{4.31}$$

Step 4

Solve the resulting equations to obtain all the parameters used in the trial solution, that is, the C_i constants.

To demonstrate how the Galerkin method can be applied to a practical problem, the displacement of a flexible cable under a uniformly distributed load is determined from the PDE. For comparison purposes, the same problem is also solved using the minimization of the *TPE*, that is, the Ritz solution.

4.5 A FLEXIBLE CABLE EXAMPLE

Consider a flexible cable of length L attached to two support points under a distributed load, $w(x)$, as shown in Fig. 4.2, where $u(x)$ is the vertical displacement of the cable below the horizontal axis. The load $w(x)$ is assumed to be uniformly distributed along the cable and is equal to w_o.

4.5.1 The trial solution for the displacement function

A polynomial approximate trial solution can be written for the displacement function. The trial solution must satisfy the essential (i.e. displacement) boundary conditions. In this problem, there are only two displacement boundary conditions at the ends of the cable:

(a) At $x = 0$, $u = 0$;
(b) At $x = L$, $u = 0$.

Also, it should be noted that the deflection of the cable is symmetrical about the midpoint ($x = L/2$), that is, the displacement slope must be zero at the midpoint. A trial solution that satisfies the essential boundary conditions and symmetry can be written as

$$u(x) = C_1 \frac{x}{L}\left(1 - \frac{x}{L}\right) \tag{4.32}$$

The slope of the cable can be obtained as the differential of the displacement function such that

$$\frac{du}{dx} = \frac{C_1}{L}\left(1 - \frac{2x}{L}\right) \tag{4.33}$$

4.5.2 Galerkin method

For a cable attached to two supports under a distributed load $w(x)$, it is possible to derive the governing partial differential equations (PDE) by considering the equilibrium of a small element of the length of the cable. By

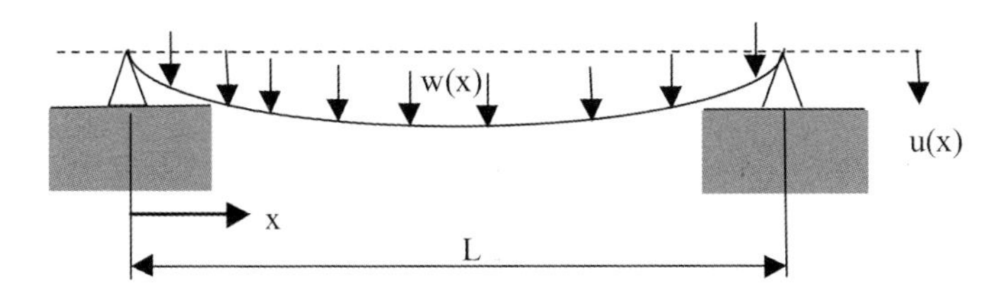

Fig. 4.2 Cable example

satisfying equilibrium, the following PDE in terms of the lateral displacement, u, can be written as

$$T\frac{\mathrm{d}^2u}{\mathrm{d}x^2}+w_\mathrm{o}=0 \tag{4.34}$$

where T is the tension in the cable. Differentiating the trial solution twice with respect to x results in

$$\frac{\mathrm{d}^2u}{\mathrm{d}x^2}=\frac{-2C_1}{L^2} \tag{4.35}$$

The Galerkin weighted residual expression can be written as

$$\int_0^L\left(T\frac{\mathrm{d}^2u}{\mathrm{d}x^2}+w_\mathrm{o}\right)\left[\frac{x}{L}\left(1-\frac{x}{L}\right)\right]\mathrm{d}x=0 \tag{4.36}$$

Substituting the displacement differentials gives

$$\int_0^L\left[T\left(\frac{-2C_1}{L^2}\right)+w_\mathrm{o}\right]\left(\frac{x}{L}-\frac{x^2}{L^2}\right)\mathrm{d}x=0 \tag{4.37}$$

By integrating from 0 to L, the value of C_1 can be obtained as

$$C_1=\frac{w_\mathrm{o}L^2}{2T} \tag{4.38}$$

Therefore, the Galerkin solution for this problem is

$$u(x)=\frac{w_\mathrm{o}L^2}{2T}\frac{x}{L}\left(1-\frac{x}{L}\right) \tag{4.39}$$

The Galerkin solution for the displacement, derived above, is identical to the exact analytical solution. This is because the chosen series solution matches the exact solution. Other combinations of the trial solutions may not give such a good agreement with the exact solution.

4.5.3 Equivalence of the Ritz and Galerkin solutions

It is important to note that although the approaches used in the Galerkin and Ritz solutions are different, the two methods are equivalent because they are effectively expressions of virtual work. To demonstrate the equivalence of the two methods, the same cable problem is now solved using the Ritz method.

For a flexible cable, stretched by a lateral load, $w(x)$, the total potential energy (TPE) can be written as

$$TPE = \int_0^L \left[\frac{1}{2} T \left(\frac{\mathrm{d}u}{\mathrm{d}x} \right)^2 - w(x)u \right] \mathrm{d}x \tag{4.40}$$

Starting from the same trial solution for the displacement in equation (4.32), and the slope, $\mathrm{d}u/\mathrm{d}x$, from equation (4.33), the following expression for TPE can be obtained.

$$TPE = \int_0^L \frac{1}{2} T \left[\frac{C_1}{L} \left(1 - \frac{2x}{L} \right) \right]^2 \mathrm{d}x - \int_0^L w_o \left[C_1 \frac{x}{L} \left(1 - \frac{x}{L} \right) \right] \mathrm{d}x \tag{4.41}$$

After integrating and algebraic manipulation, the following TPE expression in terms of C_1 is obtained.

$$TPE = \frac{1}{2} \frac{T}{3L} C_1^2 - \frac{1}{6} w_o L C_1 \tag{4.42}$$

By minimizing the TPE expression with respect to C_1, the following equation can be obtained.

$$\frac{\partial (TPE)}{\partial C_1} = 0 = \frac{T}{3L} C_1 - \frac{1}{6} w_o L \tag{4.43}$$

and the value of C_1 is determined as

$$C_1 = \frac{w_o L^2}{2T} \tag{4.44}$$

which is identical to the C_1 value obtained using the Galerkin method.

4.5.4 Advantages of the Galerkin method

The main advantages of the Galerkin method over the Ritz method are:

- It applies directly to the partial differential equations, and can be used even when an energy or a variational principle is not available or not applicable.
- The order of the differentials is one less than the variational expression. This makes the continuity requirements between elements less severe.

4.5.5 Equivalence of FE formulation and the Ritz/Galerkin methods

Finite element formulations may be regarded as special cases of the Ritz and the Galerkin methods. In the cantilever and cable problems described earlier, it is sufficient to apply the techniques to the whole geometry without having to divide it into small elements. However, this is only possible for simple geometries. In more complex problems, it becomes necessary to divide the structure into small (simple) elements.

Finite element formulation, Ritz, and Galerkin methods all use a set of trial functions to obtain approximate solutions for the displacement function. The main difference is that the FE method uses trial solutions that are defined over small elements, rather than the whole structure, and do not have to satisfy the boundary conditions. For the FE approximation to be reasonably accurate, the elements must be small in size and have a simple geometric shape. In the FE formulation, the continuity of displacements between adjacent elements has to be incorporated into the trial solutions.

4.6 SUMMARY OF KEY POINTS

- The principle of minimum total potential energy can be used in most FE formulations and is easily adaptable for different element shape functions (e.g. linear, quadratic) and different configurations (e.g. truss, beam, plate, shell).
- The Ritz method is based on using a trial function in the form of a series solution with unknown coefficients that must satisfy the boundary conditions of the problem. The principle of minimum total potential energy is then used to solve for the unknown coefficients.
- The Galerkin weighted residual approaches are widely used in the FE formulations because they operate directly on the governing partial differential equations (without the need for an energy or variational expression), and the continuity requirement between adjoining elements is less severe.
- The FE method can be seen as an equivalent method to the Ritz and Galerkin approaches, except that the FE trial solutions are defined over small elements, rather than the entire structure.

CHAPTER 5

Higher Order Quadratic Elements

The continuum elements discussed earlier assume a linear approximation of the displacements over each element. This results in a constant strain (and hence constant stress) per element. Therefore, in order to achieve good accuracy in problems where the variables change rapidly (e.g. stress concentration around a notch), a fine mesh with a very large number of linear elements should be used.

To improve the approximation within each element and to use a lesser number of elements, a 'higher order' of variation of the problem variables (here displacements) can be used by allowing the variables to change quadratically, cubically, or a higher order across the element. This is achieved by using quadratic or higher order shape functions. Owing to the added complexity of the algebraic expressions, it becomes necessary to use accurate numerical methods to perform the integrations needed to calculate the element stiffness matrix.

5.1 PROPERTIES OF THE SHAPE FUNCTIONS

In the previous analysis of continuum elements, the displacement variation with each element was expressed as a function of x and y by using a set of constants (C_1, C_2, and so on). It is more convenient and versatile to express the displacement function as an 'interpolation function', that is, in terms of the displacement values at the nodes of the element. For example, in the truss element the displacement function can be written as

$$u^*(x^*) = C_1 + C_2 x^* \tag{5.1}$$

where C_1 and C_2 are constants. The displacement conditions at the two nodes are:

- at node 1 (where $x^* = 0$), $u^* = u_1^*$;
- at node 2 (where $x^* = L_e$), $u^* = u_2^*$.

These two conditions can be used to obtain C_1 and C_2 as follows.

$$C_1 = u_1^*; \qquad C_2 = \frac{u_2^* - u_1^*}{L_e} \tag{5.2}$$

By eliminating C_1 and C_2 from equation (5.1), the displacement function can be written in terms of the nodal displacements as

$$u^*(x^*) = \left(1 - \frac{x^*}{L_e}\right)u_1^* + \left(\frac{x^*}{L_e}\right)u_2^* \tag{5.3}$$

The above equation is effectively an interpolation function for the displacement function anywhere inside the element, in terms of the two nodal displacements. The multiples of the nodal displacements are the shape functions of the element, that is,

$$u^*(x^*) = N_1(x^*)u_1^* + N_2(x^*)u_2^* \tag{5.4}$$

where N_1 and N_2 are called 'linear shape functions', and can be obtained from equation (5.3) such that

$$\begin{aligned} N_1(x^*) &= 1 - \frac{x^*}{L_e} \\ N_2(x^*) &= \frac{x^*}{L_e} \end{aligned} \tag{5.5}$$

From the above derivation for the expressions for the shape functions, it can be shown that the interpolation or shape functions must satisfy three conditions:

(a) Each shape function, N_c, must be equal to 1 at the node c itself; that is, N_1 must be equal to 1 at node 1 (when $x^* = 0$), and N_2 must be equal to 1 at node 2 (when $x^* = L_e$), as shown in Fig. 5.1.

(b) Each shape function, N_c, must be equal to 0 at nodes other than node c; that is, N_1 must be equal to 0 at node 2 (when $x^* = L_e$), and N_2 must be equal to 0 at node 1 (when $x^* = 0$).

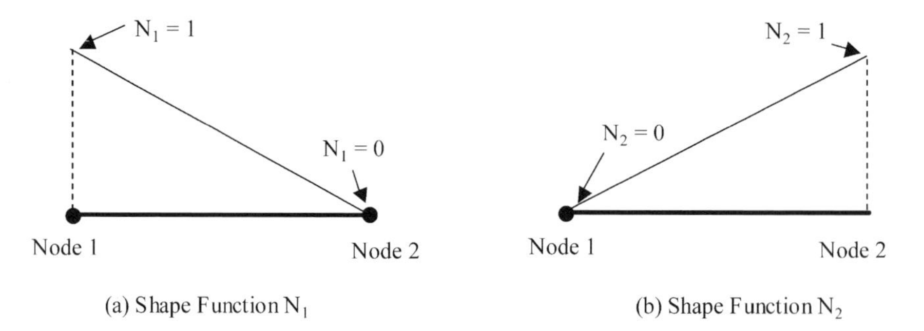

Fig. 5.1 Linear shape function variation

(c) The summation of all shape functions must add up to 1, that is, $N_1 + N_2 = 1$.

Therefore, it is possible to obtain explicit expressions for the shape functions from the above three requirements without having to use constants such as C_1, C_2, and so on. Assume a general linear expression for the shape functions

$$\begin{aligned} N_1(x^*) &= A_1 + A_2 x^* \\ N_2(x^*) &= A_3 + A_4 x^* \end{aligned} \tag{5.6}$$

where A_1, A_2, A_3, and A_4 are arbitrary constants. Applying the first shape function condition ($N_c = 1$ at node c itself) gives

$$\begin{aligned} N_1 &= 1 = A_1 + A_2(0) \\ N_2 &= 1 = A_3 + A_4(L_e) \end{aligned} \tag{5.7}$$

Applying the second shape function condition ($N_c = 0$ at nodes other than c) gives

$$\begin{aligned} N_1 &= 0 = A_1 + A_2(L_e) \\ N_2 &= 0 = A_3 + A_4(0) \end{aligned} \tag{5.8}$$

Hence we can obtain four equations from which the four unknowns (A_1, A_2, A_3, and A_4) can be obtained as

$$A_1 = 1; \qquad A_2 = \frac{-1}{L_e}; \qquad A_3 = 0; \qquad A_4 = \frac{1}{L_e} \tag{5.9}$$

which lead to the same shape function expressions of equation (5.5).

The above procedure for determining explicit expressions for the shape function can be used for any element shape and for quadratic or higher order elements. For example, linear shape functions can be written for the triangular elements as

$$\begin{aligned} u_x(x, y) &= N_1(x, y)u_{x1} + N_2(x, y)u_{x2} + N_3(x, y)u_{x3} \\ u_y(x, y) &= N_1(x, y)u_{y1} + N_2(x, y)u_{y2} + N_3(x, y)u_{y3} \end{aligned} \tag{5.10}$$

which can be expressed as a summation as

$$u_i(x, y) = \sum_{c=1}^{3} N_c(x, y)(u_i)_c \tag{5.11}$$

or as a matrix expression

$$u_i(x, y) = [N]^T [u_e] \tag{5.12}$$

where the matrix $[N]$ represents the matrix of shape functions such that

$$[N] = \begin{bmatrix} N_1 \\ N_2 \\ N_3 \end{bmatrix} \tag{5.13}$$

Note that the number of shape functions must be equal to the number of nodes in the element.

5.2 Isoparametric mapping

Although it is convenient to use the global Cartesian coordinate axes (x and y) to describe the shape functions, the algebraic expressions that evolve are very tedious and cannot be easily modified for different element shapes. Therefore, it is preferable to map the coordinate system into local coordinate systems. 'Isoparametric mapping' is defined as the method of mapping where the axes have their shapes defined by the *same* interpolation functions as the problem variables; for example, quadratic shape functions are used to describe the geometry as well as the displacement function.

Consider the eight-node isoparametric quadratic quadrilateral (IQQ) element shown in Fig. 5.2. Each side of the element is a quadratic curve with

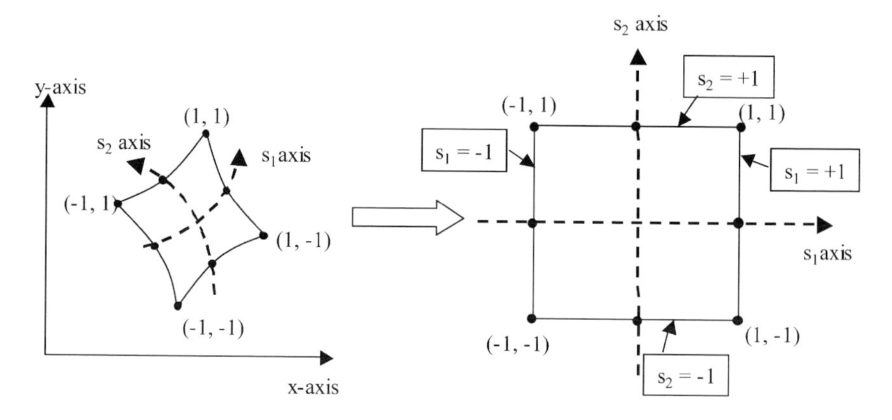

Fig. 5.2 An isoparametric quadratic (eight-node) quadrilateral element

a node at the midpoint and a node at either end. The dimensionless local axes s_1 and s_2 span from -1 to $+1$ from one side of the element to the other. The origin of the local axes is at the midpoint of the lines bisecting the sides of the quadrilateral.

Using quadratic shape functions, the geometry can be described as

$$x(s_1, s_2) = \sum_{c=1}^{8} N_c(s_1, s_2)x_c$$
$$y(s_1, s_2) = \sum_{c=1}^{8} N_c(s_1, s_2)y_c \tag{5.14}$$

Similarly, the displacement function can be expressed using the same shape functions as

$$u_x(s_1, s_2) = \sum_{c=1}^{8} N_c(s_1, s_2)(u_x)_c$$
$$u_y(s_1, s_2) = \sum_{c=1}^{8} N_c(s_1, s_2)(u_y)_c \tag{5.15}$$

or, in general

$$u_i(x, y) = [N]^T[u_e] \tag{5.16}$$

where the matrix $[N]$ represents the matrix of eight shape functions, such that

$$[N]^T = \begin{bmatrix} N_1 & N_2 & N_3 & N_4 & N_5 & N_6 & N_7 & N_8 \end{bmatrix} \tag{5.17}$$

To obtain explicit expressions for the shape functions, a general quadratic expression with eight terms is assumed for each shape function, where

$$\begin{aligned} N_c(s_1, s_2) = {} & A_1 + A_2s_1 + A_3s_2 + A_4s_1^2 + A_5s_2^2 + A_6s_1s_2 \\ & + A_7s_1s_2^2 + A_8s_1^2s_2 \end{aligned} \tag{5.18}$$

By using the properties of the shape functions, that is, $N_c = 1$ at node c itself and $N_c = 0$ at nodes other than c, as shown in Figs 5.3 and 5.4, a set of eight equations can be generated from which all the unknown constants

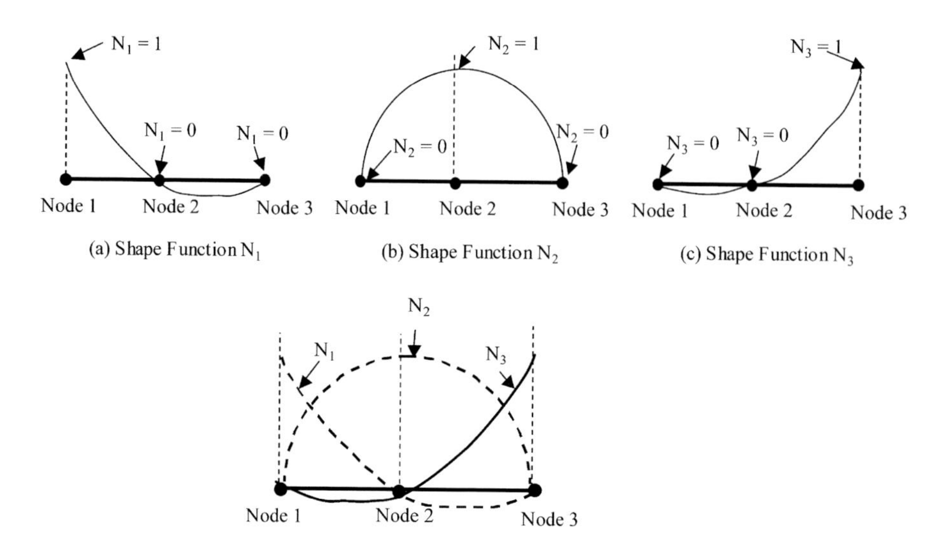

Fig. 5.3 Quadratic shape function variation over one side of a quadratic element

A_1 to A_8 can be determined. The explicit expressions for the quadratic shape functions are

$$
\begin{aligned}
N_1(s_1, s_2) &= \frac{-1}{4}(1 - s_1)(1 - s_2)(1 + s_1 + s_2) \\
N_2(s_1, s_2) &= \frac{1}{2}(1 - s_1^2)(1 - s_2) \\
N_3(s_1, s_2) &= \frac{-1}{4}(1 + s_1)(1 - s_2)(1 - s_1 + s_2) \\
N_4(s_1, s_2) &= \frac{1}{2}(1 + s_1)(1 - s_2^2) \\
N_5(s_1, s_2) &= \frac{-1}{4}(1 + s_1)(1 + s_2)(1 - s_1 - s_2) \\
N_6(s_1, s_2) &= \frac{1}{2}(1 - s_1^2)(1 + s_2) \\
N_7(s_1, s_2) &= \frac{-1}{4}(1 - s_1)(1 + s_2)(1 + s_1 - s_2) \\
N_8(s_1, s_2) &= \frac{1}{2}(1 - s_1)(1 - s_2^2)
\end{aligned}
\tag{5.19}
$$

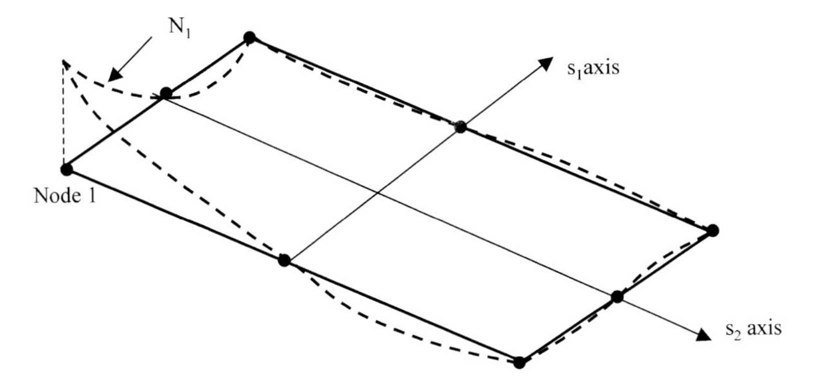

Fig. 5.4 Quadratic shape function variation of N_1 over an eight-node element

The above procedure for obtaining explicit expressions for quadratic shape functions can be adapted for any order of interpolation function and for any number of nodes. In 3D problems, the corresponding isoparametric quadratic element is a 20-node isoparametric quadratic 'brick element' with a quadratic curve representing each side of the element, as shown in Fig. 5.5. The 3D shape functions can be written as

$$u_i(s_1, s_2, s_3) = \sum_{c=1}^{20} N_c(s_1, s_2, s_3)(u_i)_c \tag{5.20}$$

5.3 Transformation of Variables (the Jacobian)

Because the shape functions are conveniently expressed in terms of the local axes (s_1, s_2), it is necessary to transform the variables from local to Cartesian (global) axes (x, y). A standard method of transformation of differentials can be used, where

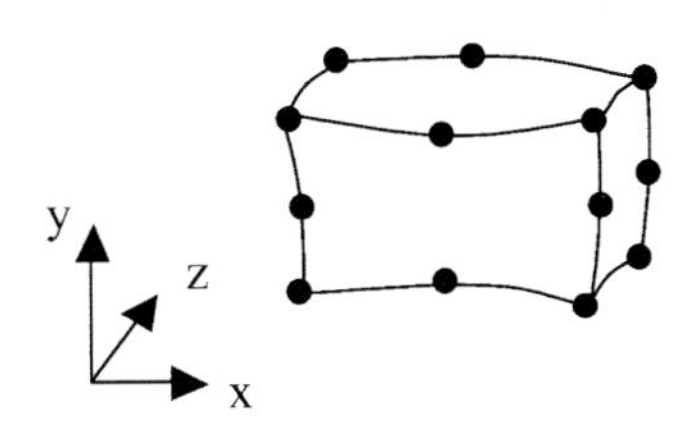

Fig. 5.5 A 20-node 3D quadratic brick element

$$\frac{\partial u}{\partial s_1} = \left(\frac{\partial u}{\partial x}\right)\left(\frac{\partial x}{\partial s_1}\right) + \left(\frac{\partial u}{\partial y}\right)\left(\frac{\partial y}{\partial s_1}\right)$$
$$\frac{\partial u}{\partial s_2} = \left(\frac{\partial u}{\partial x}\right)\left(\frac{\partial x}{\partial s_2}\right) + \left(\frac{\partial u}{\partial y}\right)\left(\frac{\partial y}{\partial s_2}\right) \tag{5.21}$$

or, in matrix form

$$\begin{bmatrix} \dfrac{\partial u}{\partial s_1} \\ \dfrac{\partial u}{\partial s_2} \end{bmatrix} = \begin{bmatrix} \dfrac{\partial x}{\partial s_1} & \dfrac{\partial y}{\partial s_1} \\ \dfrac{\partial x}{\partial s_2} & \dfrac{\partial y}{\partial s_2} \end{bmatrix} \begin{bmatrix} \dfrac{\partial u}{\partial x} \\ \dfrac{\partial u}{\partial y} \end{bmatrix} \tag{5.22}$$

The first matrix on the right-hand side is called the 'Jacobian matrix', $[J]$. Therefore, the derivatives in x and y can be written as

$$\begin{bmatrix} \dfrac{\partial u}{\partial x} \\ \dfrac{\partial u}{\partial y} \end{bmatrix} = [J]^{-1} \begin{bmatrix} \dfrac{\partial u}{\partial s_1} \\ \dfrac{\partial u}{\partial s_2} \end{bmatrix} \tag{5.23}$$

The Jacobian is also used to transform variables in integrations such that

$$\int_{\text{Area}} f(x, y)\,\mathrm{d}x\,\mathrm{d}y = \int_{-1}^{+1}\int_{-1}^{+1} f(s_1, s_2)\det[J]\,\mathrm{d}s_1\,\mathrm{d}s_2 \tag{5.24}$$

where det $[J]$ is the determinant of the Jacobian matrix $[J]$.

5.4 THE STIFFNESS MATRIX FOR HIGHER ORDER ELEMENTS

The derivation of the element stiffness matrix for higher order elements follows the same steps as those discussed earlier for linear elements. The element strains are obtained by differentiating the shape functions in equation (5.15) such that

$$\varepsilon_{xx} = \frac{\partial u_x}{\partial x} = \frac{\partial N_1}{\partial x}u_{x1} + \frac{\partial N_2}{\partial x}u_{x2} + \cdots + \frac{\partial N_8}{\partial x}u_{x8}$$
$$\varepsilon_{yy} = \frac{\partial u_y}{\partial y} = \frac{\partial N_1}{\partial y}u_{y1} + \frac{\partial N_2}{\partial y}u_{y2} + \cdots + \frac{\partial N_8}{\partial y}u_{y8}$$
$$\varepsilon_{xy} = \frac{\partial u_x}{\partial y} + \frac{\partial u_y}{\partial x} = \frac{\partial N_1}{\partial y}u_{x1} + \frac{\partial N_2}{\partial y}u_{x2} + \cdots + \frac{\partial N_1}{\partial x}u_{y1} + \frac{\partial N_2}{\partial x}u_{y2} + \cdots \tag{5.25}$$

It is convenient to re-arrange the above algebraic expressions into matrices so

$$\begin{bmatrix} \varepsilon_{xx} \\ \varepsilon_{yy} \\ \varepsilon_{xy} \end{bmatrix} = \begin{bmatrix} \frac{\partial N_1}{\partial x} & 0 & \frac{\partial N_2}{\partial x} & 0 & \cdots & \cdots & \frac{\partial N_8}{\partial x} & 0 \\ 0 & \frac{\partial N_1}{\partial y} & 0 & \frac{\partial N_2}{\partial y} & \cdots & \cdots & 0 & \frac{\partial N_8}{\partial y} \\ \frac{\partial N_1}{\partial y} & \frac{\partial N_1}{\partial x} & \frac{\partial N_2}{\partial y} & \frac{\partial N_2}{\partial x} & \cdots & \cdots & \frac{\partial N_8}{\partial y} & \frac{\partial N_8}{\partial x} \end{bmatrix} \begin{bmatrix} u_{x1} \\ u_{y1} \\ u_{x2} \\ u_{y2} \\ \cdots \\ \cdots \\ u_{x8} \\ u_{y8} \end{bmatrix} \tag{5.26}$$

or, in a more concise form

$$[\varepsilon] = [B][u_e] \tag{5.27}$$

which is similar to the expression used in the previous formulations for pin-jointed and triangular elements.

Following the same steps used previously for other elements, the element stiffness matrix can be written as

$$[k_e] = \int_v [B]^T[D][B]\,\mathrm{d}v = \int_{-1}^{+1}\int_{-1}^{+1} [B]^T[D][B]\det[J]\,\mathrm{d}s_1\,\mathrm{d}s_2 \tag{5.28}$$

The integration is performed using the Gaussian quadrature technique described below.

5.5 NUMERICAL INTEGRATION USING GAUSSIAN QUADRATURE

For simple element shapes (e.g. truss or triangular elements), it may be possible to perform some of the integrations analytically. With higher order elements, however, the integrals become too complex for analytical integration to be performed. Therefore, an efficient and accurate numerical integration scheme is required.

There are many standard numerical integration schemes that can be used to perform the FE integrations. For example, Simpson's rule is a well-known numerical integration scheme for integrating a function with a single independent variable and can be adapted for functions with two independent variables (two-dimensional integrations). Its main disadvantage, however, is that the function to be integrated is usually evaluated at fixed intervals, and is computationally inefficient if the intervals need to be small in order to maintain good accuracy. Because all element stiffnesses are calculated via an integration process, it is very important to ensure that the numerical integration procedure is as accurate as possible.

The Gaussian quadrature method (also known as Gauss–Legendre), is a popular numerical integration scheme that is computationally efficient and provides impressive accuracy. The main features of the Gaussian quadrature technique can be summarized as follows:

(a) The range of integration is from -1 to $+1$. In some cases, this may require a linear transformation of the integral variable to accommodate these limits. However, in the numerical implementation of the FE formulation, the integrals are performed over elements with shape functions that range from -1 to $+1$. Note that these integration limits have been intentionally chosen with the Gaussian quadrature technique in mind.

(b) The function within the integral range is evaluated at a given number of Gaussian points known as the 'Gaussian coordinates'. At these coordinates the function is multiplied by a weight function and added together to calculate the integral. These coordinates and weight functions are optimized to give the best possible accuracy and change according to the total number of Gaussian points used.

(c) For a given number of Gaussian points, G, the Gaussian scheme yields an exact solution up to the order $(2G - 1)$.

(d) Better accuracy is achieved with a large number of Gaussian points. However, this is only achieved at a higher cost of computation time. Therefore, an optimum number of Gaussian points must be chosen to reach a compromise between accuracy and computation time.

To demonstrate the Gaussian quadrature scheme, Fig. 5.6 shows a typical function integrated with four Gaussian points. The following equation represents the scheme for a function with a single independent variable.

$$\begin{aligned}\int_{-1}^{+1} f(x)\,\mathrm{d}x &= \sum_{g=1}^{G} f[x_g]\,w_g \\ &= f(x_1)w_1 + f(x_2)w_2 + f(x_3)w_3 + f(x_4)w_4\end{aligned} \tag{5.29}$$

where G is the total number of Gaussian points, x_g is the Gaussian coordinate, and w_g is the associated weight function.

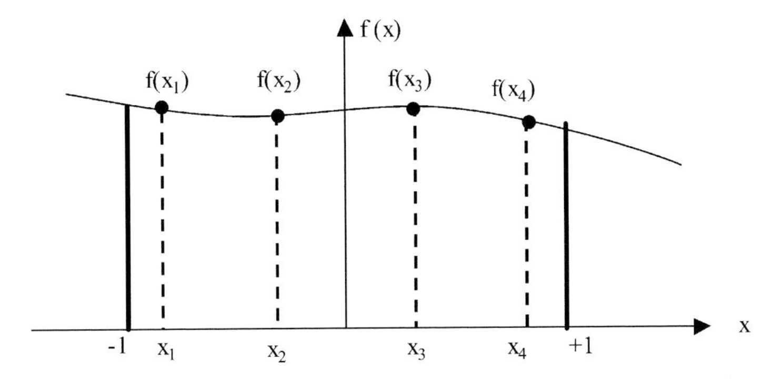

Fig. 5.6 Schematic representation of the Gaussian quadrature integration scheme

The Gaussian integration scheme can be easily adapted for functions with two independent variables (i.e. area integrals) by a nested summation such that

$$\int_{-1}^{+1}\left(\int_{-1}^{+1} f(x, y)\,\mathrm{d}x\right)\mathrm{d}y = \sum_{g_1=1}^{G_1}\left(\sum_{g_2=1}^{G_2} f[x_{g1}, y_{g2}]w_{g1}\right)w_{g2} \tag{5.30}$$

where the numbers of Gaussian points G_1 and G_2 for each summation loop may be different. Similarly, the scheme can be extended for 3D functions with three independent variables (i.e. volume integrals).

The use of Gaussian integration schemes is widespread in computer programs because of their versatility and accuracy. A comprehensive list of Gaussian formulae, coordinates, and weight functions can be found in mathematical handbooks.

To demonstrate the accuracy of the Gaussian quadrature techniques, consider the following simple integrals, which can be analytically performed such that

$$\int_{-1}^{+1}\frac{1}{(3x+5)^2}\,\mathrm{d}x = \left[\frac{-1}{3(3x+5)}\right]_{-1}^{+1} = 0.125 \tag{5.31}$$

$$\int_{-1}^{+1}\frac{x}{\sqrt{2x+3}}\,\mathrm{d}x = \left[\frac{(x-3)\sqrt{(2x+3)}}{3}\right]_{-1}^{+1} = -0.157379 \tag{5.32}$$

Using a four-point Gaussian quadrature (i.e. $G = 4$), the integrals can be expressed as a series of four terms, as shown in equation (5.29).

Table 5.1 Numerical solutions using a four-point Gaussian quadrature

Gaussian coordinate, x_g	Weight function, w_g	$f(x) = 1/(3x+5)^2$		$f(x) = x/(2x+3)^{0.5}$	
		$f(x_g)$	$f(x_g)\,w_g$	$f(x_g)$	$f(x_g)\,w_g$
−0.8611363	0.3478548	0.1712354	0.0595651	−0.7618207	−0.2650030
−0.3399810	0.6521452	0.0631279	0.0411686	−0.2232066	−0.1455631
0.8611363	0.3478548	0.0173889	0.0060488	0.3962747	0.1378461
0.3399810	0.6521452	0.0275940	0.0179953	0.1772283	0.1155786
		$\sum f(x_g)\,w_g = 0.1247778$		$\sum f(x_g)\,w_g = -0.1571415$	

Table 5.1 shows the numerical solutions using a four-point Gaussian quadrature, which are in very good agreement with the analytical solutions. This example demonstrates the high accuracy of the Gaussian integration scheme. Using more than four Gaussian points would have provided better accuracy.

5.6 SUMMARY OF KEY POINTS

- Higher order elements are elements with a quadratic or higher order variation of displacements within each element.
- Using higher order elements improves the modelling accuracy, but may consume more computation time. For a given level of accuracy, a lesser number of higher order elements is needed than linear (constant stress) elements.
- Isoparametric quadratic elements, where *both* the geometry and the displacements are allowed to vary quadratically over the element, offer the best compromise between accuracy and computation time.
- Shape functions are used to describe the variation of the geometry and displacements in the local axes of the element.
- Variables in the local element axes have to be transformed to the global (Cartesian) axes using the Jacobian of transformation.
- Owing to the added complexity of the higher order shape functions, the Gaussian quadrature technique is used to perform the numerical integrations.
- The FE formulation for higher order elements follows a similar procedure to that previously covered for truss and simple triangular elements.

CHAPTER 6

Beam, Plate, and Shell Elements

6.1 STRUCTURAL ELEMENTS

Common element types usually fall into two categories: 'continuum' and 'structural' elements, as shown in Fig. 6.1. Structural elements (such as beam, plate, and shell elements) are used when a structure exhibits bending deformation in response to external loads.

The main differences between structural and continuum elements can be summarized as follows:

- Structural elements are used to model structures in which one dimension (the thickness) is much smaller than the other dimensions. They are used in many applications to avoid having to use 3D continuum elements, which are computationally expensive.

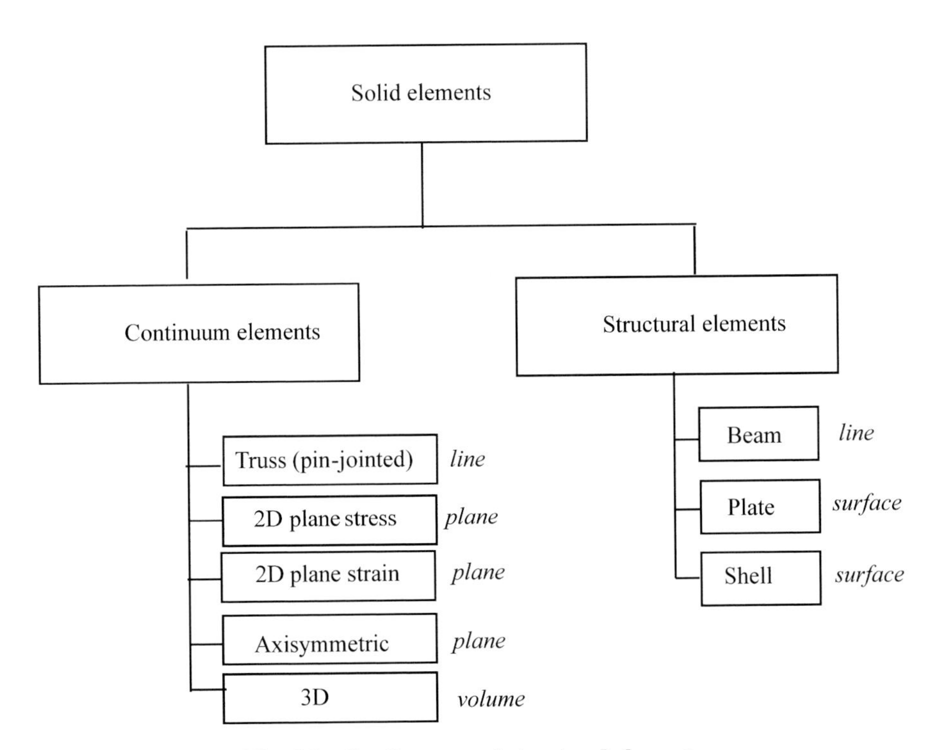

Fig. 6.1 Continuum and structural elements

- In structural elements the nodal degrees of freedom include *both* displacements and their derivatives (i.e. slopes), and are called 'Hermitian elements'. In continuum elements, the nodal degrees of freedom are *only* the displacements, and are called 'Lagrangian elements'.
- In structural elements, the rotational degrees of freedom are essential to satisfy the boundary conditions associated with bending behaviour.
- Continuum elements use a continuum elasticity approach, whereas structural elements use mechanics of materials approximations that incorporate the assumption of a small thickness.
- The FE formulation of structural elements is usually more complex than continuum elements, because displacement derivatives are used as additional nodal variables.

Figure 6.2 shows a sketch of general beam, plate, and shell elements in which the main displacements and rotations are highlighted. Before discussing the FE formulations for structural elements, it is worth highlighting the main features of these elements.

6.1.1 Beam elements

- A beam is a slender structure in which the length is significantly larger than the other two dimensions. A beam is modelled as a line with flexural stiffness. The stress in the direction of the line is considered significantly greater than the other stresses.
- In general, 3D beam elements have six degrees of freedom: three displacement components and three in-plane rotations.
- Two assumptions for the thickness can be used. In thin beams, shear deformation is ignored, and plane sections originally perpendicular to the midsection of the beam remain perpendicular to the midsurface after deformation. In thick (deep) beams, shear deformations are allowed.

6.1.2 Plate elements

- A plate is a flat surface with a small thickness normal to the middle surface, compared to other dimensions. In plates, the bending and stretching behaviours are not coupled.
- In general, 3D plate elements have three degrees of freedom: the transverse (normal) displacement and two in-plane rotations (slopes) of a line originally perpendicular to the surface.
- Two assumptions for the plate thickness can be used. In thin plates, shear deformations are ignored, and plane sections originally perpendicular to the midsection of the plate remain perpendicular to the midsurface after deformation. In thick (deep) plates, shear deformations are allowed.

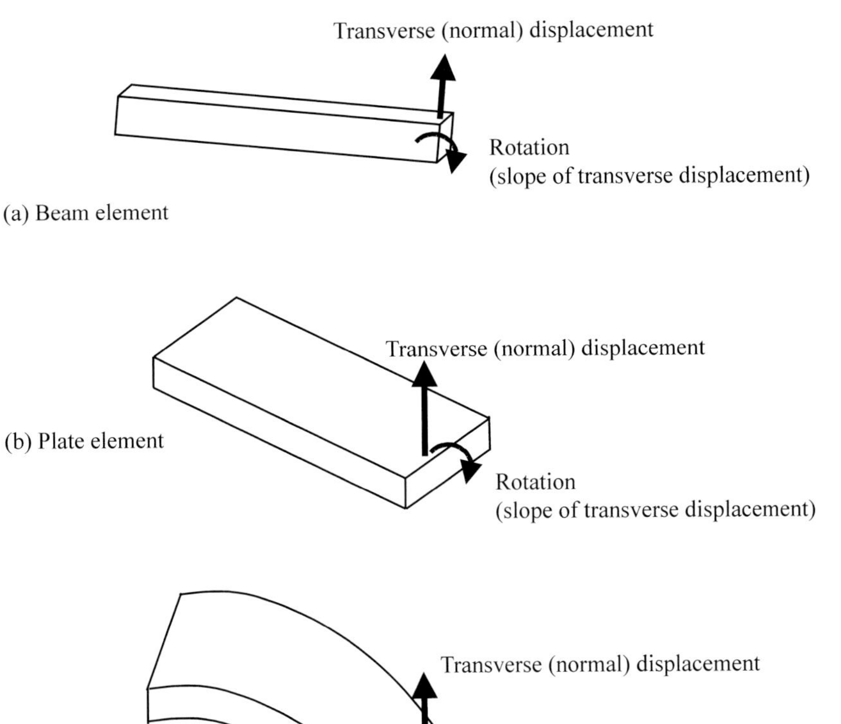

Fig. 6.2 Schematic representation of beam, plate, and shell elements

6.1.3 Shell elements

- A shell is a curved surface with a small thickness perpendicular to the middle surface, compared to other dimensions. Because the middle surface is curved, bending and stretching are coupled. Stresses in the thickness direction are neglected.
- In general, 3D shell elements have six degrees of freedom: three displacement components (two in-plane and one out-of-plane) and three in-plane rotations (slopes).
- Two assumptions for the thickness can be used. In thin shells, shear deformations are ignored, and plane sections originally perpendicular to the midsection of the shell remain perpendicular to the midsurface after

deformation. Stress differences across the shell thickness can be ignored. In thick shells, shear deformations are allowed.

- Plate and shell elements are similar. A plate element may be viewed as a flat shell. The main difference between them is that plate elements do not have mechanisms for modelling stretching behaviour. Therefore, a thin shell element can always be used to model a thin plate, but a plate element is incapable of modelling a shell element.

6.2 Beam elements

Beam elements are used to analyse frames with rigid joints where bending moments, shear forces, and axial loads are modelled. In the 2D truss elements previously used for pin-jointed structures, only one degree of freedom (the axial displacement u^*) is used. In beam elements a second degree of freedom is used: the rotation (slope), θ, at the nodes. The external force vector contains not only the shear loads, but also the bending moment applied to the beam element.

Consider a simple 2D beam element with one node at either end, as shown in Fig. 6.3. The assumptions used in deriving the FE formulation for this 2D beam element can be summarized as follows:

- Two degrees of freedom are assumed per node, the transverse (normal) displacement, u_n, perpendicular to the neutral axis, and the slope θ $(=\mathrm{d}u_n/\mathrm{d}x)$ at the nodal points.
- Points on the neutral axis (midsurface) of the beam are allowed to displace only in the perpendicular direction.
- The external loading acting on the beam is in the transverse direction acting at the neutral axis.
- The axial stress (in the direction of the beam axis) is assumed significantly greater than other stresses, that is, the transverse stress is neglected.

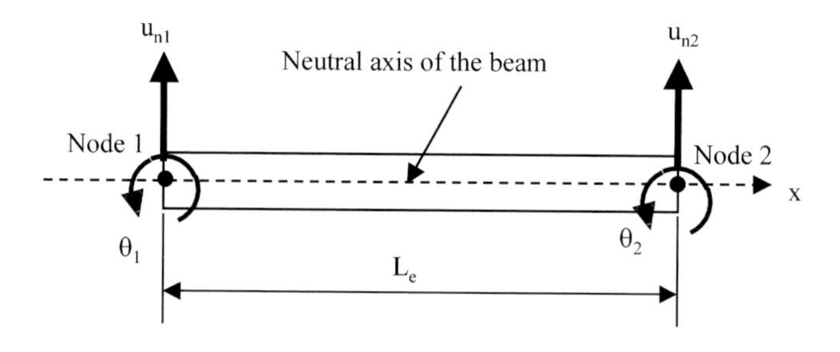

Fig. 6.3 A simple two-node beam element

The nodal variables are given by

$$[u_e] = \begin{bmatrix} u_{n1} \\ \theta_1 \\ u_{n2} \\ \theta_2 \end{bmatrix}; \qquad [F_e] = \begin{bmatrix} S_1 \\ M_1 \\ S_2 \\ M_2 \end{bmatrix} \tag{6.1}$$

where u_n is the transverse (normal) displacement and θ (=$\mathrm{d}u_n/\mathrm{d}x$) is the rotation or slope of the transverse displacement at the nodal point, x being the direction along the beam. S is the shear load and M is the bending moment.

To model a continuous beam structure, the slope at the nodal points must be continuous from one beam element to another; that is, a quadratic expression for the slope for each element is needed. Because slope is defined as the derivative of displacement, a cubic interpolation of the normal displacement over each element is needed. Note that a quadratic variation of nodal displacements over each element will result in a linear variation of slope, which is not continuous from one element to another.

Therefore, the following interpolation functions for displacement and slope are used.

$$\begin{aligned} u_n &= C_1 + C_2 x + C_3 x^2 + C_4 x^3 \\ \theta &= \frac{\mathrm{d}u_n}{\mathrm{d}x} = C_2 + 2C_3 x + 3C_4 x^2 \end{aligned} \tag{6.2}$$

where C_1 to C_4 are constants. To determine these constants, the following boundary conditions at nodes 1 and 2 must be used:

- at node 1 ($x = 0$), $u_n = u_{n1}$ and $\theta = \theta_1$;
- at node 2 ($x = L_e$), $u_n = u_{n2}$ and $\theta = \theta_2$.

This results in the following matrix expression.

$$\begin{bmatrix} u_{n1} \\ \theta_1 \\ u_{n2} \\ \theta_2 \end{bmatrix} = \begin{bmatrix} 1 & 0 & 0 & 0 \\ 0 & 1 & 0 & 0 \\ 1 & L_e & L_e^2 & L_e^3 \\ 0 & 1 & 2L_e & 3L_e^2 \end{bmatrix} \begin{bmatrix} C_1 \\ C_2 \\ C_3 \\ C_4 \end{bmatrix} \tag{6.3}$$

which can be expressed in general as

$$[u_e] = [A][C] \tag{6.4}$$

where $[u_e]$ is the displacement vector of the element, $[A]$ is the coordinate matrix, and $[C]$ contains the constants C_1 to C_4. By inverting $[A]$, the $[C]$ matrix can be obtained as

$$[C] = [A^{-1}][u_e] \tag{6.5}$$

By algebraic manipulation, the values of the constants C_1 to C_4 can be obtained as

$$\begin{aligned} C_1 &= u_{n1} \\ C_2 &= \theta_1 \\ C_3 &= \frac{3(u_{n2} - u_{n1})}{L_e^2} - \frac{\theta_2 + 2\theta_1}{L_e} \\ C_4 &= \frac{\theta_2 + \theta_1}{L_e^2} + \frac{2(u_{n1} - u_{n2})}{L_e^3} \end{aligned} \tag{6.6}$$

The FE formulation for beam elements follows the same steps as those used for continuum elements. The strain energy (SE) expression for a beam is given by

$$SE = \frac{1}{2}\int_0^{L_e} M\left(\frac{\mathrm{d}^2 u_n}{\mathrm{d}x^2}\right)\mathrm{d}x \tag{6.7}$$

where M is the bending moment at the beam cross-section defined as

$$M = EI\left(\frac{\mathrm{d}^2 u_n}{\mathrm{d}x^2}\right) \tag{6.8}$$

where E is Young's modulus and I is the second moment of area. By differentiating the expression for u_n in equation (6.2) twice with respect to x, and substituting the expression for M, the strain energy expression becomes

$$SE = \int_0^{L_e} \frac{EI}{2}(2C_3 + 6C_4x)^2\,\mathrm{d}x = \frac{EI}{2}(4C_3^2 L_e + 12C_3C_4L_e^2 + 12C_4^2L_e^3) \tag{6.9}$$

This expression can be written in a more convenient matrix form as

$$SE = \frac{EI}{2}[C]^T[Q][C] \tag{6.10}$$

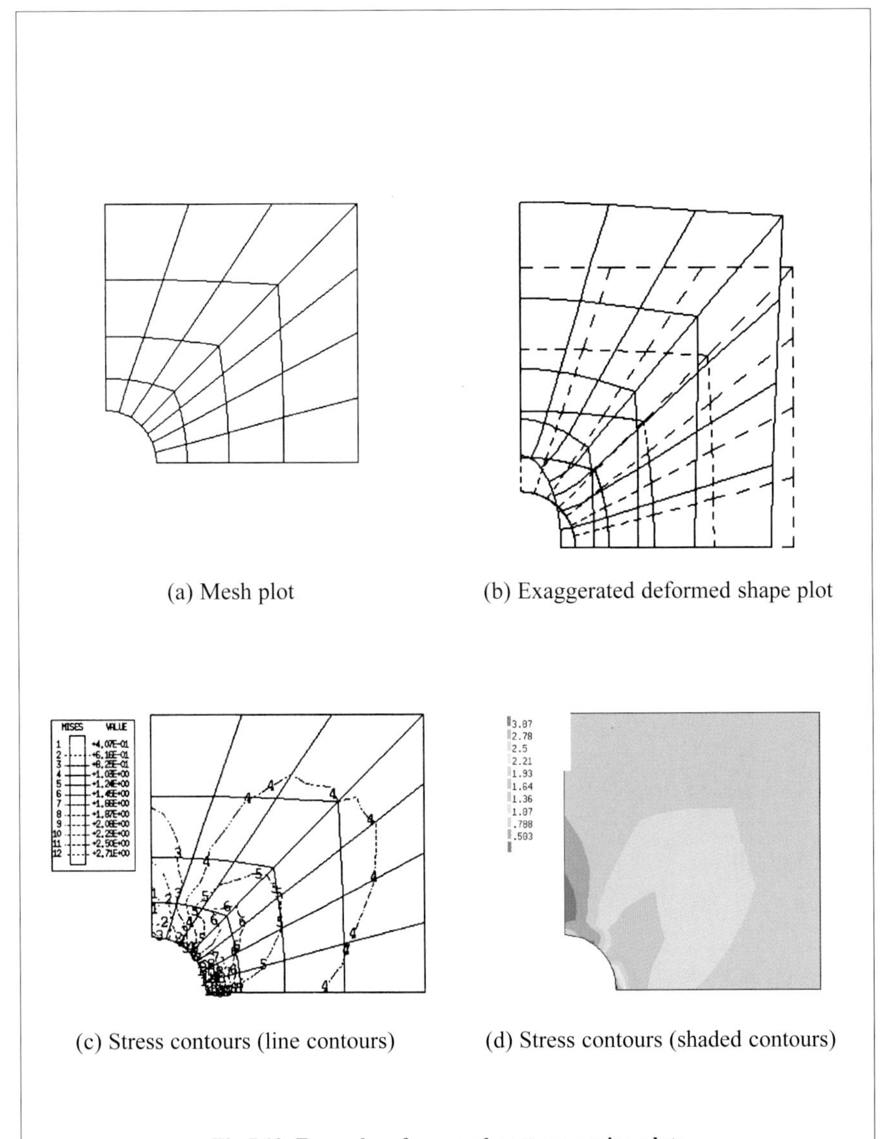

(a) Mesh plot

(b) Exaggerated deformed shape plot

(c) Stress contours (line contours)

(d) Stress contours (shaded contours)

Fig 7.10 Examples of pre- and post-processing plots

Fig 10.7 Stress contour plot (vertical stress) for the perforated plate example

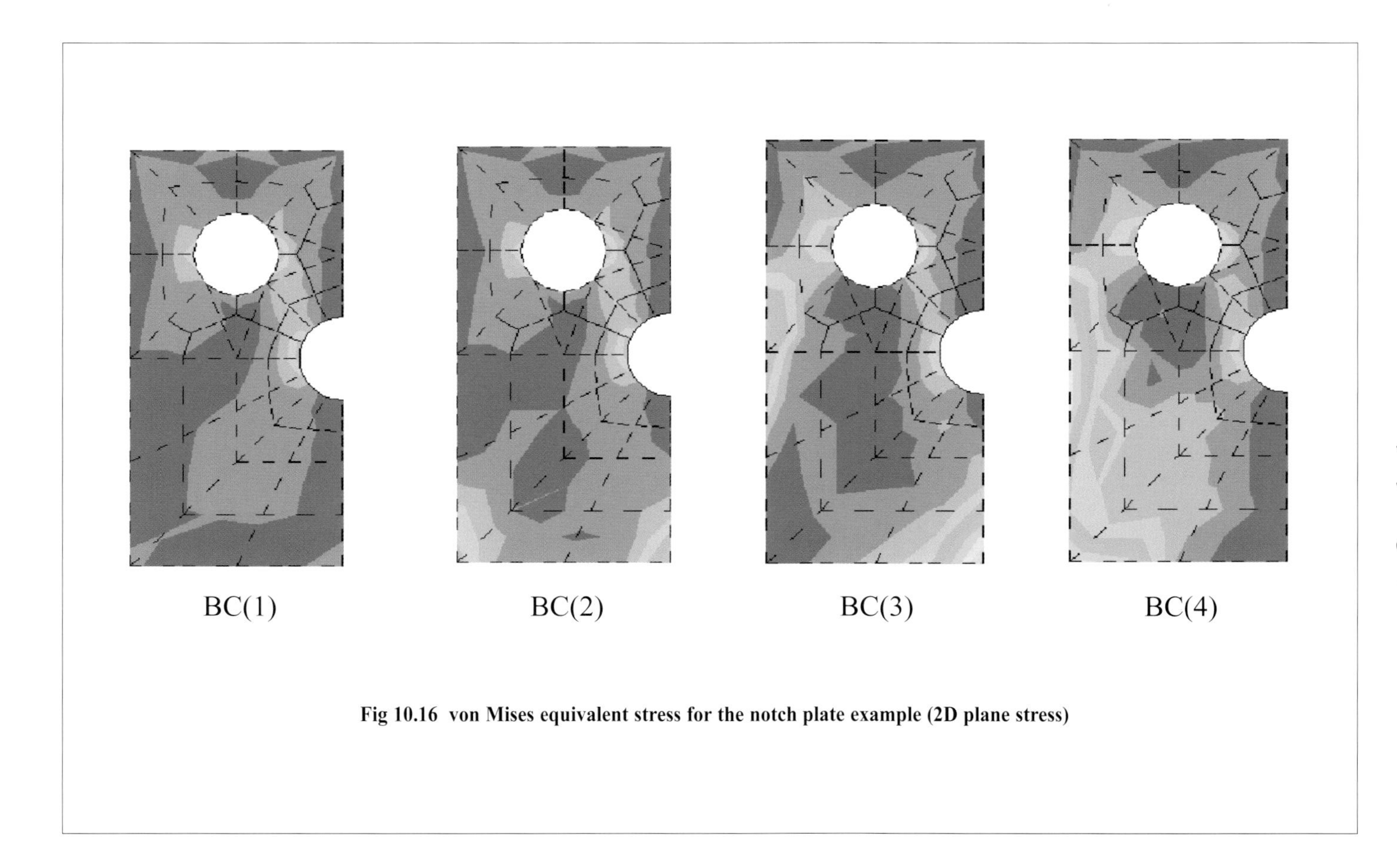

Fig 10.16 von Mises equivalent stress for the notch plate example (2D plane stress)

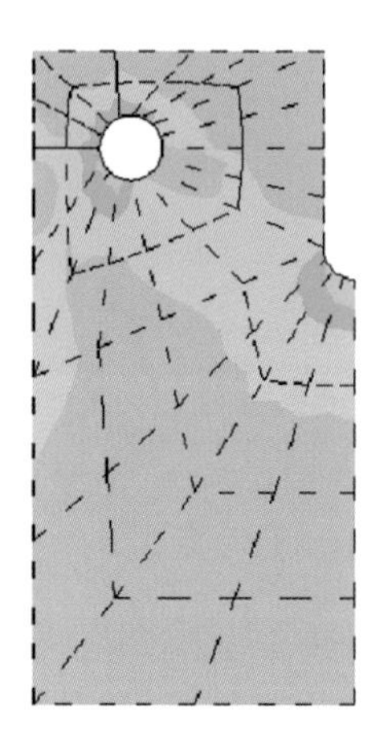

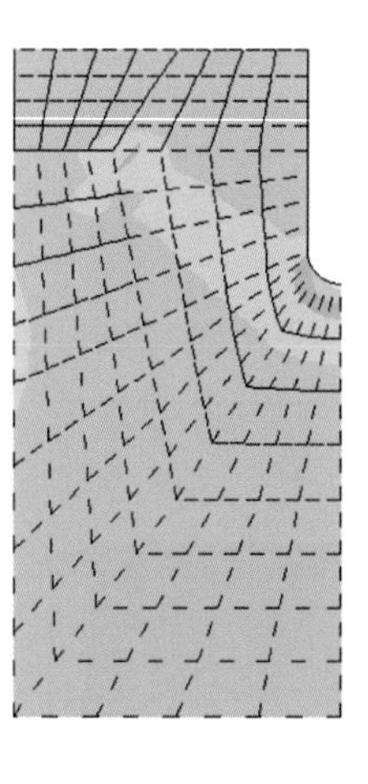

Fig 10.32 von Mises equivalent stress for the pin-loaded plate example

where $[Q]$ is defined as

$$[Q] = \begin{bmatrix} 0 & 0 & 0 & 0 \\ 0 & 0 & 0 & 0 \\ 0 & 0 & 4L_e & 6L_e^2 \\ 0 & 0 & 6L_e^2 & 12L_e^3 \end{bmatrix} \tag{6.11}$$

By substituting $[C]$ from equation (6.5), the strain energy can be expressed as a function of the displacement vector $[u_e]$, as

$$SE = \frac{EI}{2}[u_e]^T [A^{-1}]^T [Q][A^{-1}][u_e] \tag{6.12}$$

Using the expressions for the total potential energy (*TPE*) and differentiating with respect to the displacement vector, the element stiffness matrix can be obtained as

$$[k_e] = EI[A^{-1}]^T [Q][A^{-1}] \tag{6.13}$$

Substituting for $[Q]$ from equation (6.11) and performing the algebraic manipulations results in the following expression for $[k_e]$.

$$[k_e] = \left(\frac{EI}{L_e^3}\right) \begin{bmatrix} 12 & 6L_e & -12 & 6L_e \\ 6L_e & 4L_e^2 & -6L_e & 2L_e^2 \\ -12 & -6L_e & 12 & -6L_e \\ 6L_e & 2L_e^2 & -6L_e & 4L_e^2 \end{bmatrix} \tag{6.14}$$

Note that, as in previous element formulations, the stiffness matrix is symmetric.

6.3 Plate Elements

Plate elements are used to analyse flat thin structures loaded in the transverse direction. They are simpler than shell elements because they do not consider the in-plane stretching displacement function. However, they need special attention to ensure that compatibility requirements are enforced between plate elements.

For simplicity, a simple axisymmetric plate element is used to demonstrate the FE formulation for plate elements. Consider a simple two-node axisymmetric plate element of inner radius r_1 and outer radius r_2, as shown in Fig. 6.4. A transverse load per unit area is applied to the element, and S and M are the edge transverse loads and moments per unit length of the circumference, respectively.

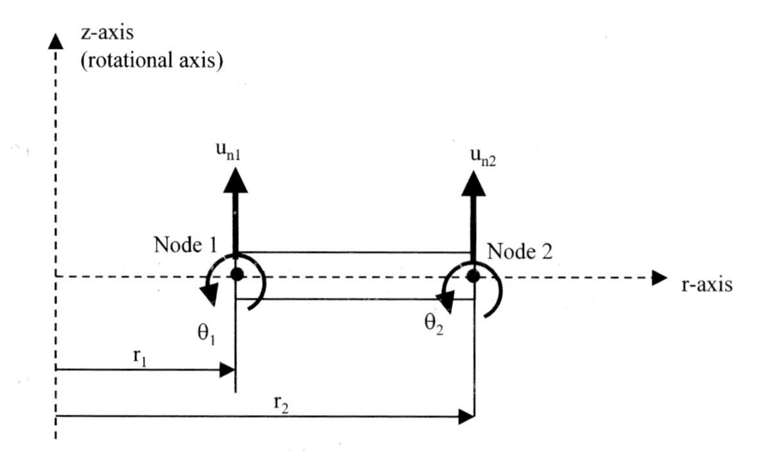

Fig. 6.4 A simple axisymmetric two-node plate element

The nodal variables for this element are the transverse (normal) displacement, u_n, and the slope, $\theta\,(=\mathrm{d}u_n/\mathrm{d}r)$, that is, two degrees of freedom at each node. As in beam elements, in order to ensure continuity between elements, a cubic variation of the displacement u_n is assumed as

$$\begin{aligned} u_n &= C_1 + C_2 r + C_3 r^2 + C_4 r^3 \\ \theta &= \frac{\mathrm{d}u_n}{\mathrm{d}r} = C_2 + 2C_3 r + 3C_4 r^2 \end{aligned} \tag{6.15}$$

where r is the radial coordinate (along the element), and C_1 to C_4 are constants. Using the coordinates of the nodal points ($r = r_1$ at node 1 and $r = r_2$ at node 2), four conditions can be used to determine the constants C_1 to C_4, as follows.

$$\begin{bmatrix} u_{n1} \\ \theta_1 \\ u_{n2} \\ \theta_2 \end{bmatrix} = \begin{bmatrix} 1 & r_1 & r_1{}^2 & r_1{}^3 \\ 0 & 1 & 2r_1 & 3r_1{}^2 \\ 1 & r_2 & r_2{}^2 & r_2{}^3 \\ 0 & 1 & 2r_2 & 3r_2{}^2 \end{bmatrix} \begin{bmatrix} C_1 \\ C_2 \\ C_3 \\ C_4 \end{bmatrix} \tag{6.16}$$

or, in general

$$[u_e] = [A][C] \tag{6.17}$$

where $[u_e]$ is the displacement vector of the element, $[A]$ is a coordinate matrix, and $[C]$ is the vector of the constants C_1 to C_4. Therefore, the $[C]$ constants can be determined as

$$[C] = [A^{-1}][u_e] \tag{6.18}$$

The strain vector in plate bending problems contains the differentials of the transverse displacements, such that

$$[\varepsilon] = \begin{bmatrix} \dfrac{\mathrm{d}^2 u_n}{\mathrm{d}r^2} \\ \dfrac{1}{r}\dfrac{\mathrm{d}u_n}{\mathrm{d}r} \end{bmatrix} \tag{6.19}$$

By differentiating the displacements in equation (6.15), the following strain expression can be derived.

$$[\varepsilon] = \begin{bmatrix} 0 & 0 & 2 & 6r \\ 0 & \dfrac{1}{r} & 2 & 3r \end{bmatrix} \begin{bmatrix} C_1 \\ C_2 \\ C_3 \\ C_4 \end{bmatrix} \tag{6.20}$$

which can be written as

$$[\varepsilon] = [X][C] = [X][A^{-1}][u_e] = [B][u_e] \tag{6.21}$$

where $[B]$ is a dimension matrix similar to that defined for other elements.
The strain energy expression for plate bending is given by

$$SE = \int_A \frac{1}{2}[M]^T[\varepsilon]\,\mathrm{d}A \tag{6.22}$$

where $[M]$ is the moment matrix defined as

$$[M] = \begin{bmatrix} M_r \\ M_\theta \end{bmatrix} = \frac{Et^3}{12(1-\nu^2)} \begin{bmatrix} 1 & \nu \\ \nu & 1 \end{bmatrix} \begin{bmatrix} \dfrac{\mathrm{d}^2 u_n}{\mathrm{d}r^2} \\ \dfrac{1}{r}\dfrac{\mathrm{d}u_n}{\mathrm{d}r} \end{bmatrix} \tag{6.23}$$

which can be written as a constitutive equation, similar to the stress–strain relationships in two-dimensional problems.

$$[M] = [D][\varepsilon] \tag{6.24}$$

where $[D]$ is the material property matrix. Following the same procedure used previously for other elements, the element stiffness matrix can be written in matrix form as

$$[k_e] = 2\pi \int_{r_1}^{r_2} [B]^T[D][B]r\,\mathrm{d}r \tag{6.25}$$

By substituting the matrices $[B]$ and $[D]$ and performing the integrations with respect to r, the following expression for $[k_e]$ can be derived.

$$[k_e] = \frac{2\pi E t^3}{12(1-\nu^2)}[A^{-1}]^T \begin{bmatrix} 0 & 0 & 0 & 0 \\ 0 & \log(r_2/r_1) & 2(1+\nu)(r_2 - r_1) & 1.5(1+2\nu)(r_2{}^2 - r_1{}^2) \\ 0 & 2(1+\nu)(r_2 - r_1) & 4(1+\nu)(r_2{}^2 - r_1{}^2) & 6(1+\nu)(r_2{}^3 - r_1{}^3) \\ 0 & 1.5(1+2\nu)(r_2{}^2 - r_1{}^2) & 6(1+\nu)(r_2{}^3 - r_1{}^3) & 1.25(5+4\nu)(r_2{}^4 - r_1{}^4) \end{bmatrix} [A^{-1}] \tag{6.26}$$

Note that the stiffness matrix is symmetric, as in other elements.

6.4 Shell Elements

Shell elements are used to model curved bodies in which the thickness of the shell is much smaller than the other dimensions. In addition to the transverse (normal) displacement incorporated in plate elements, shell elements also use the in-plane displacement as an independent variable. This allows coupling of bending and stretching actions. FE shell formulations are more complex than other structural elements.

For simplicity, a simple axisymmetric shell element is used to demonstrate the FE formulation for shell elements. Consider a simple two-node thin-shell conical element, as shown in Fig. 6.5. This type of element can be used for the analysis of axisymmetric internally pressurized thin conical shells.

The deformation of the element is assumed to be axisymmetric and dependent only on the local coordinate s along the element. The nodal variables for this element are:

(a) the in-plane meridional (stretching) tangential displacement u_t;
(b) the transverse (normal) displacement u_n; and
(c) the rotation (slope) θ $(= \mathrm{d}u_n/\mathrm{d}s)$.

Hence, this simple shell element has three degrees of freedom per node. As discussed previously for beam and plate elements, to satisfy continuity requirements, transverse displacements of the element must be approximated by a cubic polynomial expression. For simplicity, the in-plane tangential displacement, u_t, is approximated by a linear expression. Therefore, the polynomial expressions for the nodal variables are

$$\begin{aligned} u_t &= C_1 + C_2 s \\ u_n &= C_3 + C_4 s + C_5 s^2 + C_6 s^3 \\ \theta &= \frac{\mathrm{d}u_n}{\mathrm{d}s} = C_4 + 2C_5 s + 3C_6 s^2 \end{aligned} \tag{6.27}$$

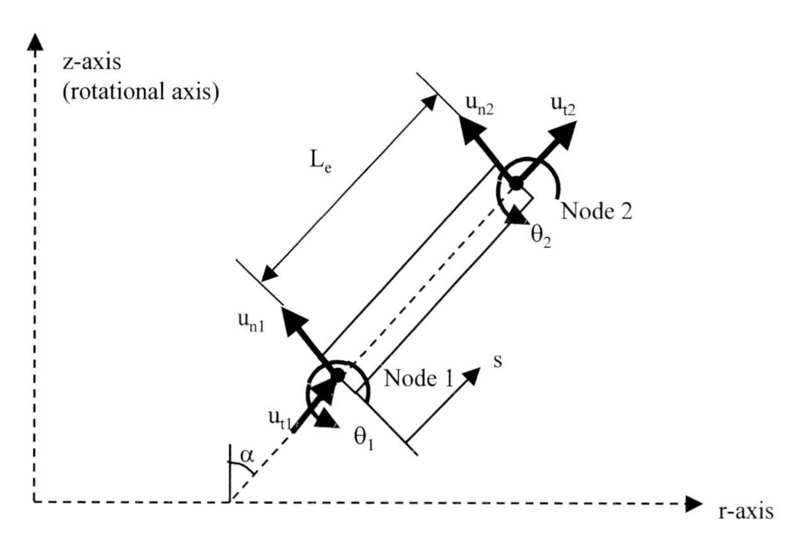

Fig. 6.5 A simple axisymmetric (conical) two-node shell element

where C_1 to C_6 are constants. Using the coordinates of the nodal points ($s = 0$ at node 1 and $s = L_e$ at node 2), six conditions can be used to determine the constants C_1 to C_6, as follows:

$$\begin{bmatrix} u_{t1} \\ u_{n1} \\ \theta_1 \\ u_{t2} \\ u_{n2} \\ \theta_2 \end{bmatrix} = \begin{bmatrix} 1 & 0 & 0 & 0 & 0 & 0 \\ 0 & 0 & 1 & 0 & 0 & 0 \\ 0 & 0 & 0 & 1 & 0 & 0 \\ 1 & L_e & 0 & 0 & 0 & 0 \\ 0 & 0 & 1 & L_e & L_e^2 & L_e^3 \\ 0 & 0 & 0 & 1 & 2L_e & 3L_e^2 \end{bmatrix} \begin{bmatrix} C_1 \\ C_2 \\ C_3 \\ C_4 \\ C_5 \\ C_6 \end{bmatrix} \tag{6.28}$$

or, in general

$$[u_e] = [A][C] \tag{6.29}$$

where $[u_e]$ is the displacement vector of the element, $[A]$ is the coordinate matrix, and $[C]$ is the vector of the constants C_1 to C_6. The $[C]$ constants can be determined as

$$[C] = [A^{-1}][u_e] \tag{6.30}$$

In thin shell theory, the strain vector contains the extensional strains as well as curvature changes, and is defined as

$$[\varepsilon] = \begin{bmatrix} \dfrac{\mathrm{d}u_t}{\mathrm{d}s} \\ \dfrac{1}{r}(u_t \sin\alpha + u_n \cos\alpha) \\ \dfrac{\mathrm{d}^2 u_n}{\mathrm{d}s^2} \\ \dfrac{1}{r}\left(\sin\alpha \dfrac{\mathrm{d}u_n}{\mathrm{d}s}\right) \end{bmatrix} \tag{6.31}$$

where α is the inclination of the element, as shown in Fig. 6.5.

By differentiating the displacements in equation (6.27), the following strain expression is obtained.

$$[\varepsilon] = \begin{bmatrix} 0 & 1 & 0 & 0 & 0 & 0 \\ \dfrac{\sin\alpha}{r} & \dfrac{s\sin\alpha}{r} & \dfrac{\cos\alpha}{r} & \dfrac{s\cos\alpha}{r} & \dfrac{s^2\cos\alpha}{r} & \dfrac{s^3\cos\alpha}{r} \\ 0 & 0 & 0 & 0 & 2 & 6s \\ 0 & 0 & 0 & \dfrac{\sin\alpha}{r} & \dfrac{2s\sin\alpha}{r} & \dfrac{3s^2\sin\alpha}{r} \end{bmatrix} \begin{bmatrix} C_1 \\ C_2 \\ C_3 \\ C_4 \\ C_5 \\ C_6 \end{bmatrix} \tag{6.32}$$

which can be written as

$$[\varepsilon] = [X][C] = [X][A^{-1}][u_e] = [B][u_e] \tag{6.33}$$

where $[B]$ is a dimension matrix similar to that defined for other elements.

The strain energy expression for the shell problems is

$$SE = \int_V \frac{1}{2}[\sigma]^T[\varepsilon]\,\mathrm{d}V \tag{6.34}$$

where V is the volume. Following the same procedure used previously for other elements, the element stiffness matrix can be written in matrix form as

$$[k_e] = 2\pi \int_{L_e} [B]^T[D][B]r\,\mathrm{d}s \tag{6.35}$$

6.5 SUMMARY OF KEY POINTS

- Structural elements such as beam, plate, and shell elements are used to model structures in which one dimension (the thickness) is much smaller than the others.
- Structural elements use displacement as well as rotation (i.e. displacement slope) as the degrees of freedom.
- Simple beam elements in two dimensions have two degrees of freedom per node, the transverse (normal) deflection, u_n, and the slope, θ $(=\mathrm{d}u_n/\mathrm{d}x)$ where x is the direction along the element. The load vector contains shear loads and bending moments.
- Plate elements may be regarded as flat shell elements in which the bending and stretching actions are uncoupled.
- Simple plate elements have two degrees of freedom per node, the transverse (normal) displacement, u_n, and the slope, θ $(=\mathrm{d}u_n/\mathrm{d}r)$, where r is the direction along the element.
- In shell elements, since the structure is initially curved, the bending and stretching actions are coupled.
- Simple axisymmetric shell elements have three degrees of freedom per node, in-plane tangential displacement, u_t, transverse (normal) displacement, u_n, and slope, θ $(=\mathrm{d}u_n/\mathrm{d}s)$, where s is the direction along the element.
- More sophisticated higher order structural elements can be used with higher order shape functions.

CHAPTER 7

Practical Guidelines for FE Applications

7.1 INTRODUCTION

Some practical guidelines on how to use the FE method to model engineering problems are presented in this chapter. Using FE software to solve engineering problems is not an exact science, and often requires the application of good engineering judgement. There are many issues with which inexperienced users of FE often struggle, including:

- choosing the correct element type (e.g. 2D plane stress, plane strain, three-dimensional, beam, shell, and so on);
- choosing the best element order (e.g. linear, quadratic, and so on);
- deciding on the level of mesh refinement;
- simulating the real-life boundary conditions; and
- assessing the accuracy of the FE solutions in the absence of other (non-FE) solutions for comparison.

This chapter covers an outline of the data input requirements for FE analysis, good and bad practice, solution accuracy, and commercial FE codes.

7.2 DATA INPUT FOR FE CODES

To model a given problem using FE software, the user must specify, without ambiguity, all of the data required to define a problem with a unique solution. These include:

- geometry;
- material properties;
- analysis type;
- displacement boundary conditions;
- applied loads;
- element type; and
- other information, such as the objective of the analysis.

It is essential to define all of the above data before attempting to use FE software. A typical 'FE data input sheet' is shown in Fig. 7.1. It is advisable

	Example	*Explanation*
REF. NO.	Test001 (file name Test1.inp)	• Specify a unique reference number (and the file name containing the data input)
DESCRIPTIVE TITLE	Perforated plate analysis (symmetric quarter)	• Specify a unique descriptive title for the problem
GEOMETRY	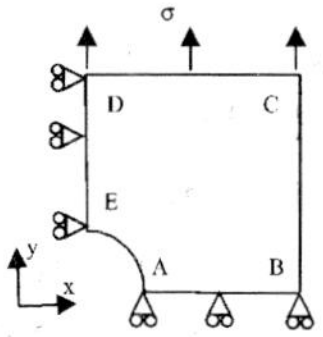 2D Plane stress Continuum elements Length of CD = 100 mm Hole radius = 20 mm Applied stress, $\sigma = 100\ \mathrm{N/mm^2}$	• Sketch the geometry • Show all displacement constraints • Show the applied loads • Specify the dimensionality (i.e. 2D plane strain/stress, 3D, axisymmetric, etc.) • Specify the configuration (e.g. continuum, beam, shell, plate) • Specify units (even if not used in the analysis)
MATERIAL PROPERTIES	$E = 250.0 \times 10^3\ \mathrm{N/mm^2}$ $\nu = 0.25$ Plasticity model: isotropic hardening, with hardening modulus = $50.0 \times 10^3\ \mathrm{N/mm^2}$	• Specify all material properties relevant to the analysis • Specify all units used (this is particularly important for nonlinear problems) • Specify relevant material law (e.g. creep or plasticity law)
ANALYSIS TYPE	Static elastic–plastic analysis (nonlinear material law)	• Specify relevant analysis required (e.g. elastic/plastic/creep, thermal, static/dynamic, linear/nonlinear, etc.)
DISPLACEMENT BOUNDARY CONDITIONS	Zero y-displacement (roller conditions) specified on line AB Zero x-displacement (roller conditions) specified on line DE	• Specify the displacement constraints (referring to the sketch shown in the GEOMETRY section)
APPLIED LOADS	A uniform tensile stress (distributed load) specified at the top surface (line CD)	• Specify all the applied loads and their units (referring to the sketch shown in the GEOMETRY section)

Fig. 7.1 A typical FE data input sheet

ELEMENT TYPE	Eight-node isoparametric quadratic element with 2 × 2 Gauss integration points	• Specify the type of element used and the number of integration points (e.g. element code used in the FE software)
OTHER INFORMATION	Objective: to determine the stress concentration around the hole	• State any relevant information regarding the FE model, such as: – objective of the analysis – any special features, e.g. load applied in a number of load steps, initial conditions, etc.

Fig. 7.1 Continued

to use a data input sheet for each new FE run. This not only helps the analyst in record-keeping, but also helps to explain, in a concise manner, the data used in the FE software to other analysts.

7.3 Accuracy and convergence of FE solutions

7.3.1 Sources of error in FE analysis

Finite element solutions are not exact solutions – they involve many approximations and assumptions. There are three main sources of error that may occur in FE solutions:

(a) *Modelling errors*, which occur if the geometry is not exactly modelled or the boundary conditions are not accurately interpreted.
(b) *Mesh errors*, which may occur if the mesh is not a 'good' mesh, for example, containing long thin elements, not sufficiently refined in regions of sharp variation of variables, and so on.
(c) *Numerical errors*, which may occur due to round-off in the computations, where numbers are truncated due to insufficient digits being used in the calculations. In some problems, the solution matrix may become 'ill-conditioned', that is, very sensitive to small changes in the variables. This can occur, for example, when the stiffness matrix contains coefficients of varying orders of magnitudes in the same row, due to a large variation in element sizes or modulus of elasticity between elements. Numerical errors can also occur in the numerical integration procedures.

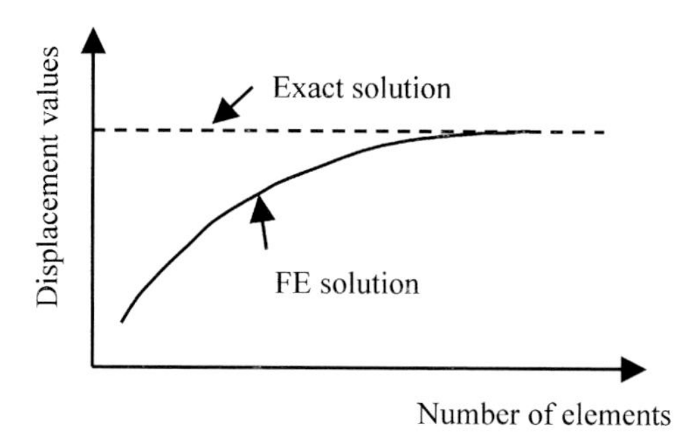

Fig. 7.2 Convergence of FE solutions

7.3.2 Convergence of FE solutions

For a given problem, as the FE mesh is refined, the FE solution should approach the exact solution, in other words, converge. This should not be confused with convergence in nonlinear problems, where iterations are used.

The displacement-based FE formulation usually gives an overestimate of the true stiffness of the element; that is, elements are assumed 'over-stiff'. Therefore, since stiffness multiplies the displacements to obtain the external forces, the *displacements are underestimated.* Because stresses are calculated from the displacement values, this means that *stresses are also underestimated.* In other words, the errors in the displacement values (and stresses) are always negative, as shown in Fig. 7.2. This, however, assumes that a reasonably good mesh is used in the FE model. If a poor mesh is used, then it is possible that some nodal displacement values will be overestimated while most others will be underestimated. This is an important design consideration when using FE solutions to arrive at stress values.

7.4 General guidelines for using FE software

Finite element programs should not be used as 'black boxes' without a firm understanding of the underlying theory and principles behind the technique. The results should always be examined closely for discrepancies or checked against other solutions, whenever possible.

Some general guidelines for using FE software to analyse engineering problems are given below:

- *Choose the correct element type.* The elements chosen for the problem must be of the correct geometrical type (configuration) with valid assumptions, for example, 2D plane strain/plane stress, axisymmetric,

3D, plate, beam, shell, and so on. The most appropriate element order should be chosen (e.g. linear, quadratic). In general, isoparametric quadratic elements are suitable for most problems. The elements must fit the surface or boundary of the problem. If the geometry has awkward boundary shapes, higher order elements, for example, quadratic elements, should be used.

- *Use a good mesh.* A 'good' FE mesh may be defined as the mesh with the minimum number of elements required to arrive at an acceptable solution accuracy. Unnecessary use of very fine meshes is wasteful of computer resources, and may be seen as an indication of a poor understanding of the FE model. In regions of expected sharp stress gradients, for example, at a notch or near a hole, the elements must be of small size, and gradually increase in size as the distance from this region increases. This is similar to curve fitting, where more points are needed to fit a rapidly varying curve. Element sizes must not change abruptly from one element to another; that is, element sizes must change gradually between adjacent elements. This is necessary to ensure that the degree of approximation is evenly spread across the FE mesh.
- *Avoid long thin elements.* The 'aspect ratio' of the element, defined as the ratio of the largest to the smallest side, must be as close to 1 as possible; that is, similar in shape to a square, an equilateral triangle, or a cube. Long thin elements always cause errors, but can be tolerated if there are only a few of them and they are placed in regions of low stress gradients. Finite element programs usually perform a check on the aspect ratio of the elements to identify any 'bad' elements, that is, elements with a high aspect ratio.
- *Perform a mesh convergence study.* It is often difficult to establish the optimum mesh refinement needed for an FE model. One way of checking that the FE mesh used is a reasonable one is to start from a relatively coarse mesh, then refine it (e.g. by doubling the number of elements) to see the effect on the solutions. If the solutions exhibit large changes, this indicates that further mesh refinement may be necessary. However, it is important to note that this is not a fool-proof approach, because a badly designed FE mesh will always give inaccurate solutions, regardless of the mesh refinement used.
- *Check stress accuracy.* Because displacements are the primary unknowns, displacement values are always more accurate than the stress values. If stress values are desired to a high degree of accuracy, a fine mesh should be used. Stress values are always more accurate at the Gauss integration points, rather than the averaged nodal stress value (which is usually obtained by extrapolation and interpolation from the

Gauss points). Discontinuities in stress values between adjacent elements (before averaging) should be checked. These discontinuities indicate the error in the approximations and should be small, for example, within 5 percent of the element stress.

- *Prevent rigid body motion.* 'Rigid body motion' means the body may move as a whole without straining. This violates the 'small deformation' assumptions used in the FE formulation. Therefore, rigid body motion must be prevented by specifying displacement constraints in all Cartesian coordinate directions. This is discussed further in the next section.
- *Check reaction forces.* Check that the summation of the externally applied loads is equal, or approximately equal, to the summation of the reaction forces at the constrained nodes.
- *Use a benchmark.* Before attempting a new analysis, use a benchmark, that is, a problem for which a reliable solution exists. A benchmark can be used to test the accuracy of the FE software, and to learn how to correctly use the FE software. Benchmarks are particularly important in nonlinear problems.
- *Ensure inter-element connectivity (if different types of elements are used).* If element types are mixed within the same FE mesh, for example triangular elements adjoin quadrilateral elements, then it is important to ensure that adjoining elements are correctly interconnected such that there are no holes or gaps. This ensures that the compatibility relationships between adjacent elements are not violated. The best solutions are obtained when all the elements in the FE mesh have the same number of nodes and the same order of variation (e.g. all linear or all quadratic). As a general rule, adjacent element sides should have the same number of nodes and the same order of variation.
- *Check the text output of the FE software.* After a successful FE run, the text output should be checked for warnings, for example, about bad element shapes, nodes with duplicate or conflicting boundary conditions, or prescribed loads, problems in the solver, and so on.

7.5 Preventing Rigid Body Motion

Linear FE formulations are based on the assumption that element displacements and strains are small. If a structure is insufficiently restrained, it may move as a whole, that is, undergo rigid body motion, which would invalidate the FE solutions. Consider, as an example, a 2D problem involving a notched steel plate, under a stress of, say, 100 MPa, as shown in Fig. 7.3(a). Because the forces are in equilibrium, the summation of all forces in the

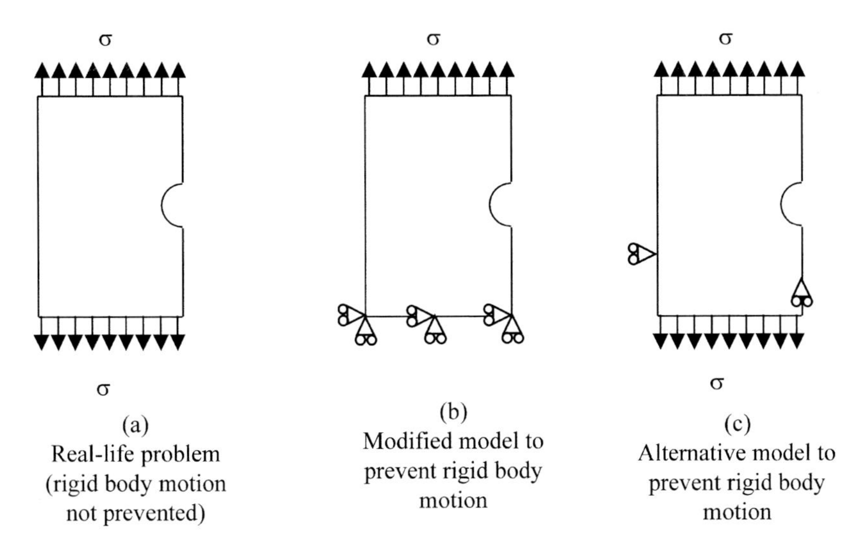

Fig. 7.3 Preventing rigid body motion

x- and y-directions must be zero. This would be true if an exact analytical solution is derived. However, because FE solutions are approximate due to the round-off error in the computational operations, the summation of all the nodal forces will never be exactly equal to zero, but equal to a very small negligible number, say 10^{-10} N in the y-direction. Because this is a nonzero force, it will be sufficient to cause the body to move as a whole in the y-direction, thus invalidating the small deformation assumption.

One way of preventing this from occurring is to fix the bottom surface, that is, cement it to a rigid surface, as shown in Fig. 7.3(b). This would make the FE solution valid because the small negligible 'residual' force error, say 10^{-10} N, would only cause a negligible displacement in the constrained body. However, the resulting deformed shape of the FE model would be different from that of the real-life problem. Despite this, the stress concentration at the notch would be the same, or almost the same, provided that the notch is sufficiently far away from the bottom surface. Another possibility of imposing sufficient displacement constraints is shown in Fig. 7.3(c).

In many FE problems, it becomes necessary for the user to introduce additional displacement constraints to prevent rigid body motion. The location of the constraint must be chosen to be as far away as possible from the region of interest (usually the region of stress concentration). This is effectively an application of the 'St. Venant principle', which states that if a structure is subjected to two statically equivalent load cases, then the stresses

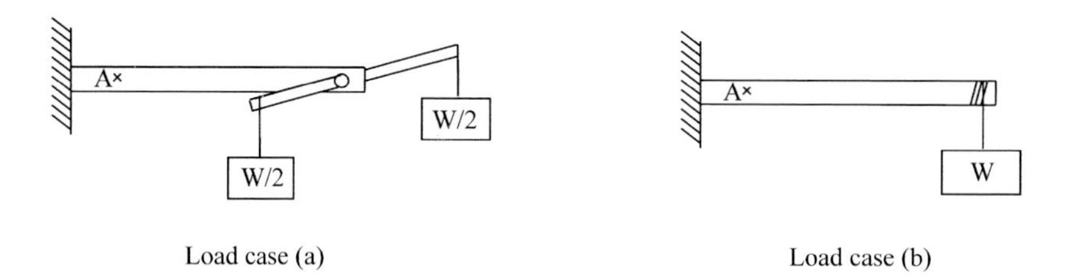

Fig. 7.4 Example of the application of St. Venant's principle

and displacements remote from the point of application of the load are unaffected by the details of the load application. As an example, consider the cantilever problem shown in Fig. 7.4, where two load cases are considered. In load case (a), two loads each of magnitude $W/2$ are applied through a rigid pin inserted in the beam. In load case (b), a hanging weight W is attached to the free end of the beam. Because both load cases are statically equivalent, the stress and strains at point A, assuming it is far away from the load, are the same regardless of which load case is used. However, at or near the loaded end of the beam, the stresses and strains are very different in load cases (a) and (b).

Therefore, it is very important to prevent rigid body motion by imposing sufficient displacement constraints on the structure. The displacement constraints must be placed in all Cartesian directions. For example, in a 2D analysis, if there is a displacement constraint in the x-direction but no displacement constraint in the y-direction, the body will move as a rigid body in the y-direction. If it becomes necessary to add additional artificial displacement constraints to prevent rigid body motion, then they should be placed away from the region of interest so that the overall deformations and stresses in the model are not adversely affected.

7.6 Examples of Good and Bad Practice

Figure 7.5 shows a summary of some good and bad practices in applying the FE method.

7.7 Stress Accuracy and Integration Points

In engineering analysis, stresses are usually of more interest than displacements. In FE solutions they are calculated per element from the

Good Practice	Bad Practice
Using good element shapes (close in shape to an equilateral triangular, a square or a cube) 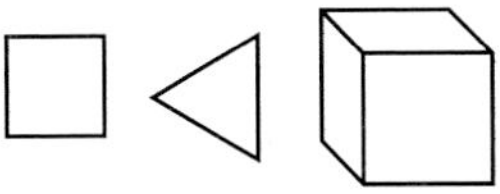	Using long thin elements (with large aspect ratio) 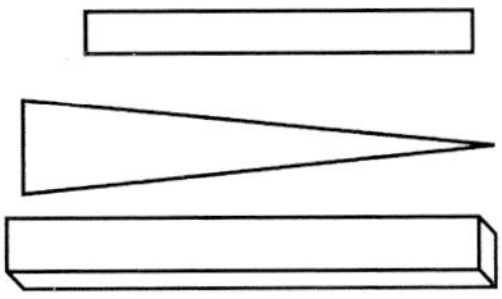
Using a well-graded mesh (with a gradual, not abrupt, change in size of adjacent elements) 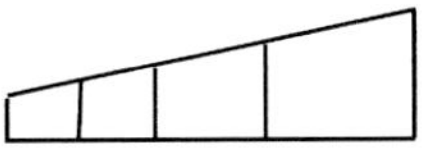	Using a mesh with abrupt changes in the size of adjacent elements 
Using quadratic (or higher-order) elements to fit a circular or curved boundary 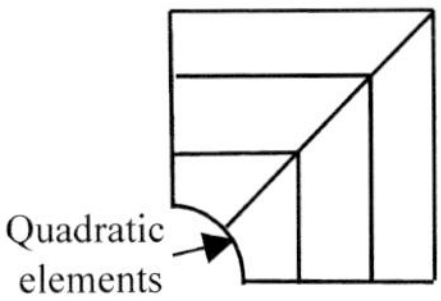 Quadratic elements	Using straight-sided (linear) elements to fit a circular or curved boundary 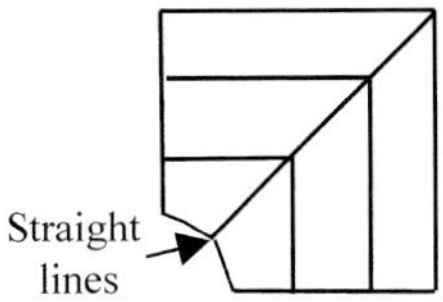 Straight lines
Preventing rigid body motion (in both x- and y-directions) 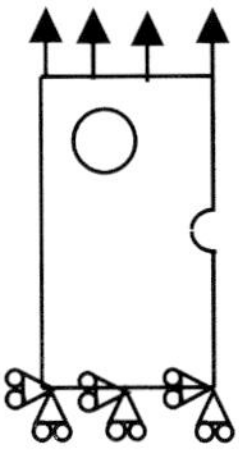	Rigid body motion not prevented (here additional displacement constraints are needed) 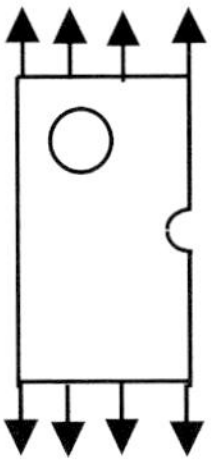

Fig. 7.5 Examples of good and bad practice

Taking advantage of symmetry to reduce problem size (here a quarter of the geometry is modelled)

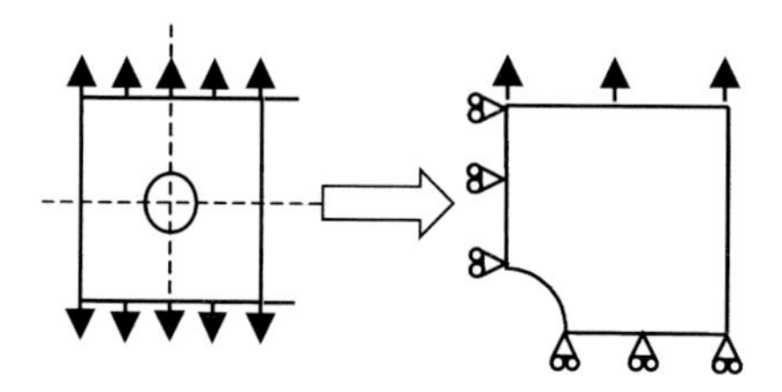

Symmetry not used

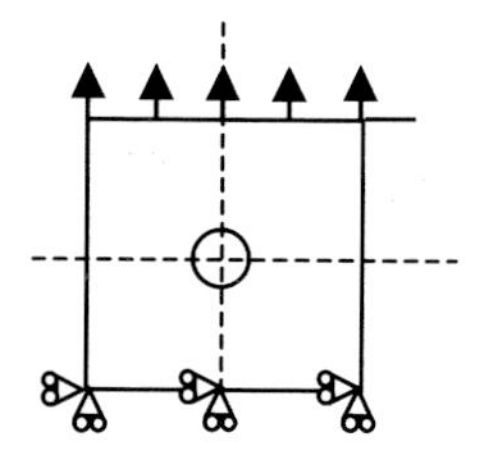

Using adjacent element sides with the same number of nodes and the same order of variation

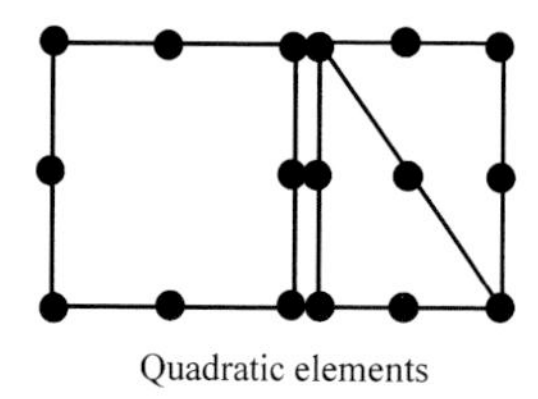

Mixing linear and quadratic elements where adjacent element sides have different order of variation

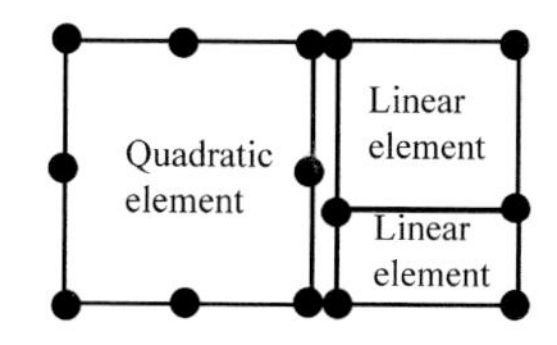

Corner nodes connected to other corner nodes

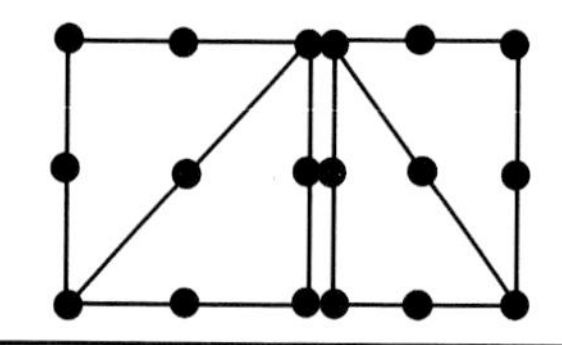

Corner nodes connected to mid-side nodes

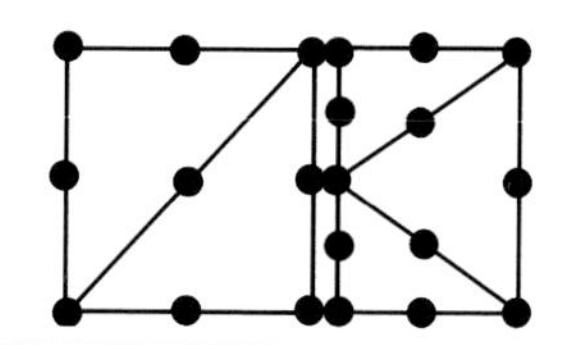

Fig. 7.5 Continued

nodal displacement values. The nodal displacements are differentiated to obtain the element strains, and then substituted into the material law (i.e. the stress–strain relationships) to obtain the element stresses.

7.7.1 Average nodal stress values

Element and nodal stresses are calculated by a numerical integration procedure using the Gaussian quadrature technique. The stresses are computed at the integration points (also called Gauss points), which are located inside the elements (not at the nodes of the element).

Most FE programs use one of several approaches to evaluate the stress at a given node, as follows:

- Stresses are interpolated and extrapolated from the Gauss integration points of the element.
- The stresses are averaged at a node by taking into consideration all the elements connected at that node.
- For a given element type, 'good stress points' can be identified. Interpolation and extrapolation can then be used from the stress values at these points.

7.7.2 Full and reduced integration

Better accuracy is obtained as the number of integration points is increased. For example, for an eight-node isoparametric quadratic quadrilateral element, shown in Fig. 7.6, a 3×3 integration scheme (i.e. nine integration points) is sufficient for good accuracy. This is called 'full integration'. A lesser number than nine Gaussian points for eight-noded elements, for example, 2×2 points, is usually referred to as 'reduced integration'. Therefore, for each element type it is possible to use either full or reduced integration.

Using full integration is obviously numerically more accurate than using reduced integration. However, in view of the fact that the displacement-based FE formulations always overestimate the element stiffness (i.e. the element is over-stiff), it has been established that using a lesser number of integration points produces a less stiff element. Therefore, using a less accurate method of integration (i.e. reduced integration) tends to produce FE solutions closer to the real-life behaviour of structures. However, using reduced integration indiscriminately can be risky because, in certain applications, instability may occur due to the element stiffness approaching zero (this is called the 'hour glass mode').

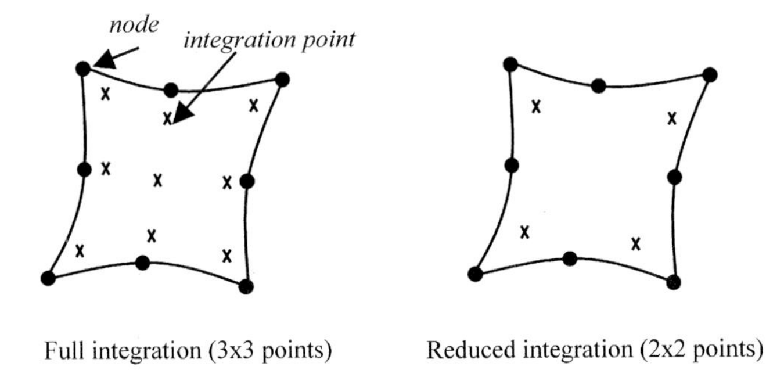

Fig. 7.6 Examples of full and reduced integration

Reduced integration is usually recommended for problems involving material nonlinearities such as plasticity and creep problems, or incompressible materials, such as rubber deformation.

7.7.3 Equivalent nodal forces

Applied loads may be prescribed as distributed loads (i.e. stresses) or concentrated point loads. To simulate a distributed load, such as a constant pressure applied to one edge of the element, the element shape functions must be taken into consideration to arrive at kinematically equivalent nodal forces. This is not straightforward, because the distribution of the forces is not obtained simply by dividing the total load by the number of nodes on the element edge, but by integrating the distributed load over the element edge, using the shape functions, to arrive at the equivalent forces at each node.

Figure 7.7 shows examples of the equivalent nodal forces for linear and quadratic elements that should be applied to simulate a constant pressure, *p*,

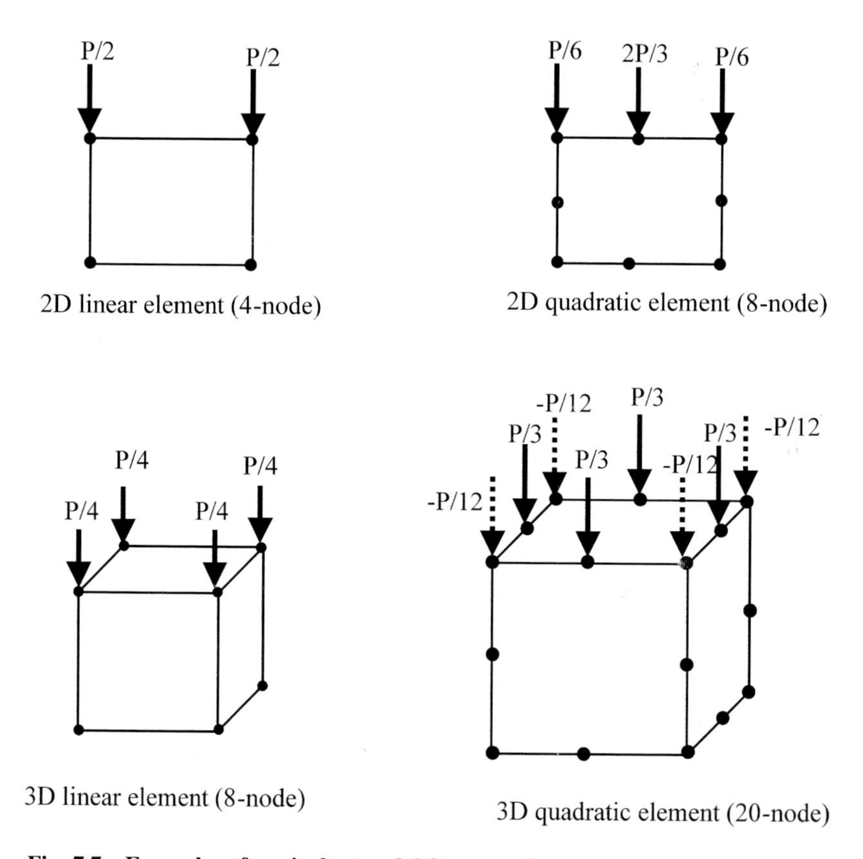

Fig. 7.7 Examples of equivalent nodal forces on linear and quadratic elements

applied to the element edge. Note that in quadratic elements, the midside node has a higher proportion of the load than the edge nodes. Also, it is interesting to note that in 3D quadratic 20-node brick elements, some of the corner equivalent forces are negative, that is, tensile. This can cause problems in problems involving contact analysis, where tensile nodal forces may be assumed to indicate the separation of the contacting nodes.

7.8 THE PATCH TEST FOR CONVERGENCE

The patch test is a simple test used to indicate the potential performance of a given element. A group (patch) of elements is subjected to nodal displacements (as boundary conditions on the boundary of the patch) such that a state of constant strain (hence stress) is obtained everywhere within the patch.

Using a plane stress FE analysis, these nodal displacements are prescribed on the boundary of a patch of elements of arbitrary orientation, as shown in Fig. 7.8. If the computed displacements at the internal nodes correspond to the prescribed displacements and the computed element strains and stresses are constant regardless of the element orientation, then the patch test is successfully passed.

For example, consider a square plate (plane stress) subjected to a prescribed displacement at each of the boundary nodes such that

$$\begin{aligned} u_x &= A_1 + A_2 x + A_3 y \\ u_y &= A_4 + A_5 x + A_6 y \end{aligned} \tag{7.1}$$

By differentiating the displacement, the element strains can be obtained as

$$\begin{aligned} \varepsilon_{xx} &= A_2 \\ \varepsilon_{yy} &= A_6 \\ \varepsilon_{xy} &= A_3 + A_5 \end{aligned} \tag{7.2}$$

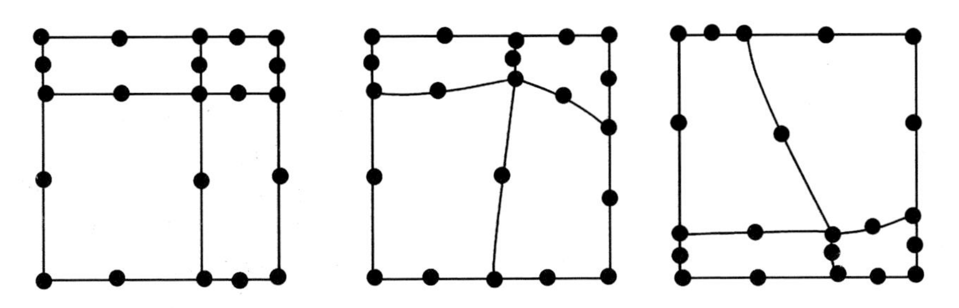

Fig. 7.8 The patch test for elements

If a variety of element orientations and sizes are used, and give correct results, then the patch test is passed for this type of element.

7.9 Commercial FE software packages

A large number of commercial FE packages are available for engineering analysis. Most commercial FE software packages are available on a wide range of computer hardware systems ranging from personal computers to large mainframe computers. The capabilities of the FE software vary widely, but most general-purpose FE commercial codes offer the following features:

- Analysis capabilities, e.g. stress analysis, heat conduction, dynamic behaviour, and so on;
- Element library, e.g. 2D, 3D, axisymmetric, beam, plate, shell, quadratic, and higher order elements;
- Material behaviour models, e.g. elastic, elasto-plastic, creep, visco-elastic, anisotropic, and so on;
- Automatic mesh generation to minimize the amount of labour required to generate a mesh;
- A range of boundary conditions and loads, e.g. point constraints, applied loads, pressures, heat flux, contact with friction; and
- Graphical pre- and post-processing facilities, e.g. to plot deformed shapes, stress contours, graphs of variables, and so on.

Finite element analysis can be broken down into three distinct stages (Fig. 7.9):

- *Stage 1:* The *pre-processor*, in which the mesh is generated and the data input file is constructed;
- *Stage 2:* The *solver*, in which the matrices are assembled and solved ('number crunching' stage);
- *Stage 3:* The *post-processor*, which is used to view the results of the solver.

Stage 1: Pre-processing

The pre-processing stage produces an input file that contains all the data needed to define the problem. The most time-consuming part of generating an input file for an FE analysis is the mesh generation, that is, the specification of the coordinates of each node and each element in the mesh. Commercial FE packages usually offer automatic or semi-automatic mesh generation, for example, generating the mesh by simply specifying a few points on the boundary and the number of elements. This also helps to avoid

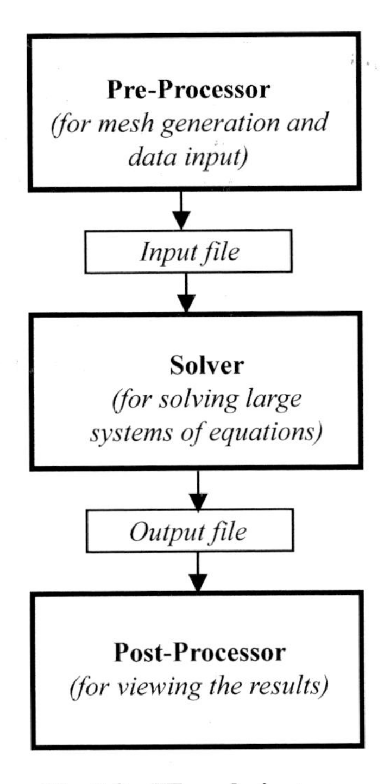

Fig. 7.9 FE analysis stages

mistakes in the mesh description. Finite element meshes can also be generated from CAD drawings or 3D solid models.

A good pre-processor should be able to detect obvious errors in the data input, for example, identifying long thin elements, incorrectly specified elements, missing information, and so on.

Pre-processors are usually interactive and use colour graphics to aid the visualization of the mesh. They also exhibit good visibility features such as zooming, rotating, changing the angle of view, removing hidden lines, and so on, which are particularly useful in generating 3D meshes.

Stage 2: The solver

The input file generated by the pre-processor is used to construct the element stiffness matrices of each element, and then assemble the overall system of simultaneous equations. The nodal displacement vectors are first calculated, and then used to obtain the strains, stresses, and other variables. The solver stage is often referred to as 'number crunching', because of the very large number of mathematical operations involved in constructing and solving the overall system of equations. After a successful run, the solver should produce an 'output file', which contains all relevant information, such as displacements, stresses, strains, temperatures, and so on at all nodes and elements. The output file can be a text file, which can be directly viewed by the user, or a binary file, which is intended for the post-processor.

Stage 3: Post-processing

Post-processing is used for examining the computed values, such as displacements or stresses at specific locations. Commercial FE software packages usually offer the user an interactive coloured graphical display of the results as well as printed listings of the variables. Features of post-processors include the following:

- *Deformed shapes*: These are produced by multiplying the deformations of all the nodes by a suitable magnification factor and adding them to the nodal coordinates. This produces an 'exaggerated' deformed shape, which may be used as a qualitative observation of the general trend of the displacements, as well as checking that the displacement constraints have been correctly described. However, it is important to remember that exaggerated deformed shapes may be misleading.
- *Stress contours*: These are produced by extrapolation and interpolation of the stress values to generate 'colour contours' of constant stress (similar to those used in geographical maps). They provide a useful visual interpretation of the results, for example, highlighting stress concentration areas. Most FE software codes allow the user to manipulate the contours, for example, to show only certain levels of stress values. Although FE solutions produce some degree of stress discontinuity between elements, stress contours are often 'smoothed'; that is, presented as a gradual blend of colours.
- *X–Y charts*: These are used to plot the variation of any variable along a given path, such as the variation of a stress component on a specific line or the variation of stress with time in time-dependent problems. These plots are very useful for inclusion in reports.
- *Vector plots*: Vectors can be plotted to represent the magnitude and direction of some variables, such as principal stresses.

Some examples of pre- and post-processing plots are shown in Fig. 7.10.

7.10 SUMMARY OF KEY POINTS

- FE solutions should always be carefully checked, and not taken for granted to be accurate.

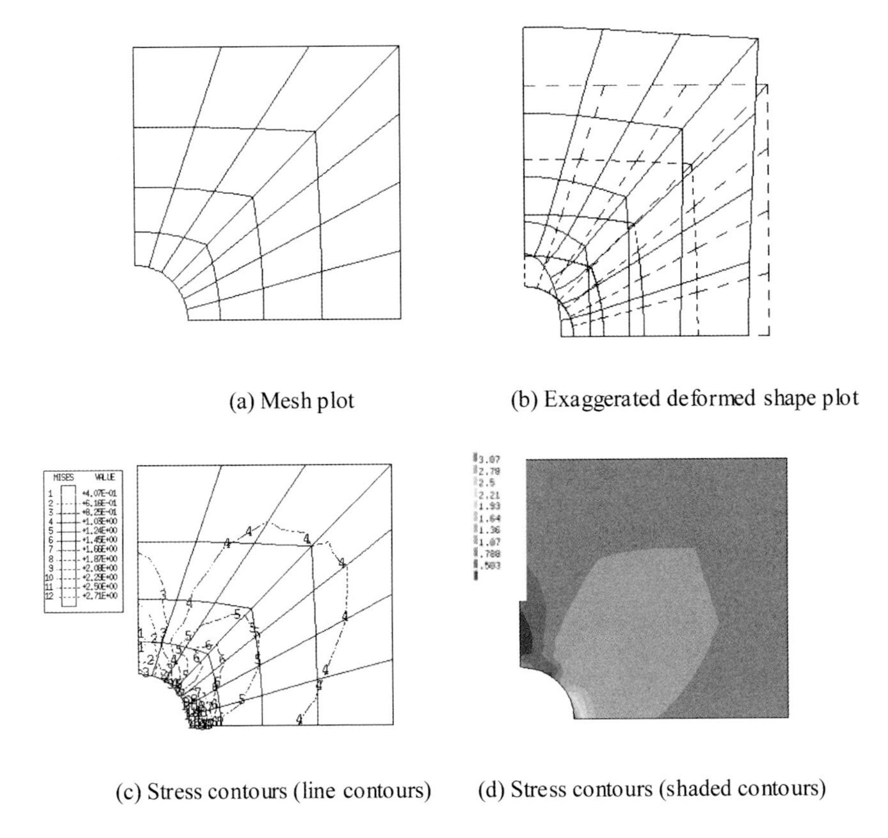

(a) Mesh plot (b) Exaggerated deformed shape plot

(c) Stress contours (line contours) (d) Stress contours (shaded contours)

Fig. 7.10 Examples of pre- and post-processing plots
(See colour plate section.)

- Displacements and stresses are usually underestimated in FE analysis.
- For best accuracy, the FE mesh should be refined in regions of rapidly changing stresses and should not contain long and thin elements.
- Rigid body motion must be prevented by ensuring that there are sufficient displacement constraints in all Cartesian directions.
- The patch test can be used to indicate the potential performance of a given element.
- Computed stress values are less accurate then the displacement values. Full integration can improve the stress accuracy, but requires more computation time than reduced integration.
- Pre- and post-processing facilities are very useful in checking the FE solutions and visual interpretation of the solutions.

CHAPTER 8

Introduction to Nonlinear FE Analysis

8.1 INTRODUCTION

In linear analysis, the behaviour of the structure is assumed to be completely reversible; that is, the body returns to its original undeformed state upon the removal of the applied loads, and solutions for various load cases can be superimposed. In many engineering applications, however, the behaviour of the structure may depend on the load history or may result in large deformations beyond the elastic limit. For such nonlinear problems, the solutions from several load cases cannot be superimposed.

Examples of nonlinear applications include elastoplasticity of metals, creep behaviour, buckling, metal forming, and contact problems. In this chapter, a description of the main three types of nonlinearities is presented, that is, material nonlinearity, geometric nonlinearity, and boundary nonlinearity (contact).

The application of the FE method to nonlinear problems usually requires the use of small load increments and/or an iterative procedure. Iterations are usually performed to ensure that the solution is convergent, that is, the error in approximating the equilibrium state is acceptably small. An outline of the general nonlinear FE procedures is presented.

Solving nonlinear problems using FE techniques requires much more effort than linear problems, with many additional issues that need to be addressed, such as convergence, automatic load incrementation, accuracy, and so on. The main difficulties that may be encountered by the FE user and the FE code are discussed.

8.2 COMPARISON OF LINEAR AND NONLINEAR PROBLEMS

In linear analysis, the displacements are assumed to be linearly dependent on the applied loads, and the behaviour of the structure is assumed to be completely reversible. This means that solutions for various load cases can be superimposed. In many engineering applications, however, the relationships between the displacements and the applied loads are nonlinear. For such problems, the user must carefully consider how to approach the analysis, model the problem using FE, and use the results of the analysis.

The main differences between linear and nonlinear analysis are summarized in Table 8.1.

Table 8.1 Comparison of linear and nonlinear problems

Feature	*Linear problems*	*Nonlinear problems*
Load–displacement relationship	Displacements are assumed to be linearly dependent on the applied loads.	The load–displacement relationships are usually nonlinear.
Stress–strain relationship	A linear relationship is assumed between stress and strain.	In problems involving material nonlinearity, the stress–strain relationship is a nonlinear function of stress, strain, and/or time.
Magnitude of displacement	Changes in geometry due to displacement are assumed to be small and hence ignored, and the original (undeformed) state is always used as the reference state.	Displacements may not be small, hence an updated reference state may be needed.
Material properties	Linear elastic material properties are usually easy to obtain.	Nonlinear material properties are difficult to obtain and may require additional experimental testing.
Units	Units of material properties and applied loads have to be consistent, and can be scaled.	Units of material properties and applied loads are very important, and cannot be scaled.
Reversibility	The behaviour of the structure is completely reversible upon removal of the external loads.	Upon removal of the external loads, the final state may be different from the initial state.
Superposition	Solutions for various load cases can be linearly superimposed.	Solutions from several load cases cannot be superimposed.
Loading sequence	Loading sequence is not important, and the final state is unaffected by the load history.	The behaviour of the structure may depend on the load history, hence the load may have to be applied sequentially.
Iterations and increments	The load is applied in one go with no iterations.	The load is often divided into small increments with iterations performed to ensure that equilibrium is satisfied at every load increment.
Computation time	Computation time is relatively small in comparison to nonlinear problems.	Owing to the many solution steps required for load incrementation and iterations, computation time is high, particularly if a high degree of accuracy is sought.

Table 8.1 Continued

Feature	*Linear problems*	*Nonlinear problems*
Robustness of solutions	A solution can easily be obtained with no interaction from the user.	In difficult nonlinear problems, the FE code may fail to converge without some interaction from the user.
Initial state of stress/strain	The initial state of stress and/ or strain is unimportant.	The initial state of stress and/ or strain is usually required for material nonlinearity problems.

8.3 CLASSIFICATIONS OF NONLINEAR PROBLEMS

Traditionally, nonlinear problems in solid mechanics have been grouped into three main types: material, geometric, and boundary nonlinearities.

8.3.1 Material nonlinearity

Here the stress–strain constitutive relationships are nonlinear. Such material nonlinearities are usually further classified into three categories:

(a) Time-independent behaviour such the elasto-plastic behaviour in metals in which the material is loaded past the yield point.
(b) Time-dependent behaviour such as creep of metals at high temperatures in which a power law stress–strain relationship is often used and the effect of variation of stress/strain with time is studied.
(c) Viscoelastic/viscoplastic behaviour in which both the effects of plasticity and creep are exhibited. Here the stress is dependent on the strain rate.

Figure 8.1 shows a typical example of elasto-plastic behaviour in a uniaxial test specimen when subjected to a tensile force in a tensile testing machine. Up to the yield point, the linear assumption can be used, but after the yield point a significant error occurs if linear theory is used, because the stress–strain relationship becomes nonlinear.

8.3.2 Geometric nonlinearity

Geometric nonlinearity occurs when the changes in the geometry of a structure due to its displacement under load are taken into account in analysing its behaviour. The change in geometry affects both equilibrium and kinematic, that is, strain–displacement, relationships. In geometric nonlinearity, the equilibrium equations take into account the deformed shape,

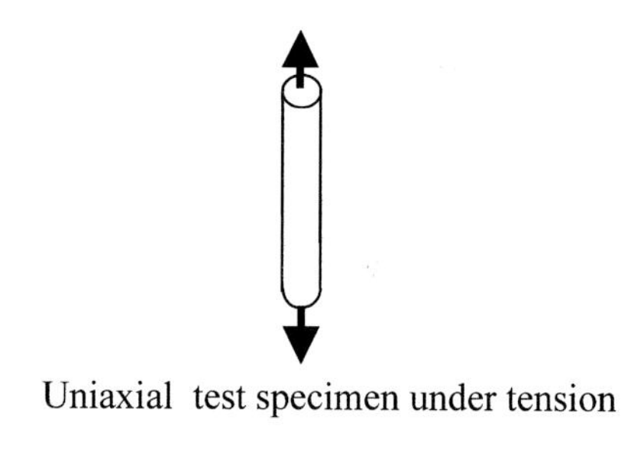

Uniaxial test specimen under tension

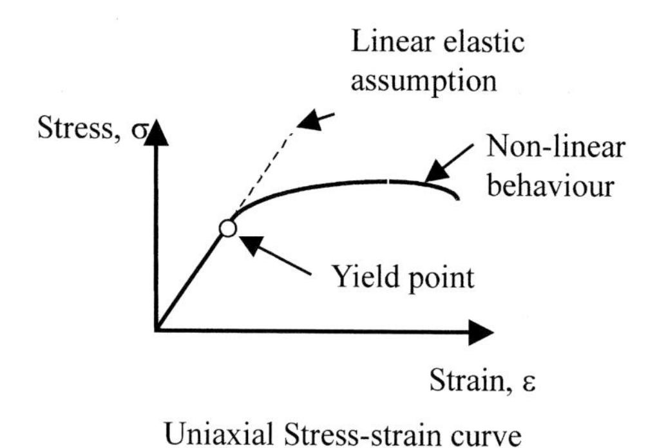

Uniaxial Stress-strain curve

Fig. 8.1 Example of a material nonlinearity problem

whereas in linear analysis the equilibrium equations are always based on the original (undeformed) shape. As a consequence of this, the strain–displacement relationships may have to be redefined to take into account the current (updated) deformed shape.

It should be emphasized that large and small displacements, strains, or rotations are possible; that is, geometric nonlinearity does not always mean 'large' displacements. In many thin plate structures, for example, displacements of the order of the plate thickness will give rise to geometric nonlinearities. Moreover, large displacements are not always associated with large strains.

Examples of geometric nonlinearities with small strains and small rotations include the deformation of shallow shells and arches subjected to lateral loading and elastic buckling of struts, whereas the displacement of a fishing rod under the weight of a heavy fish is a typical example of small strains with large rotations. More complex geometric nonlinearity may involve large displacements, large rotations, and large strains, and also changes in the loading and boundary conditions as the structure deforms.

Figure 8.2(a) shows an example of nonlinear geometric behaviour in a cantilever beam under a transverse load. At large loads, the linear assumption

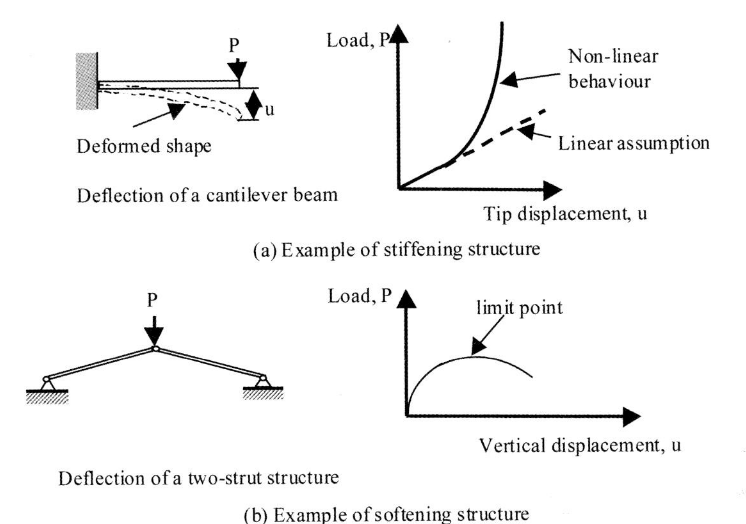

Fig. 8.2 Examples of geometric nonlinearity problems

is inadequate because it predicts a proportionally increasing tip displacement, whereas in real-life applications, after a certain tip displacement, the cantilever begins to stiffen and further displacement is hindered. An example of geometric nonlinearity in a two-bar system in which the structure softens is shown in Fig. 8.2(b). As the downward displacement is increased, a 'limit point' is reached in the load–displacement curve after which the load has to drop to maintain equilibrium.

8.3.3 Boundary nonlinearity (contact)

Boundary nonlinearity occurs in most contact problems, in which two surfaces come into or out of contact, and the displacements and stresses of the contacting bodies are not linearly dependent on the applied loads. This type of nonlinearity may occur even if the material behaviour is assumed linear and the displacements are small, due to the fact that the size of the contact area is not linearly dependent on the applied loads; that is, doubling the applied loads does not necessarily produce double the contact area or double the displacements. If the effect of friction is included in the analysis, then a stick–slip behaviour may occur in the contact area, which adds a further nonlinear complexity that is normally dependent on the loading history.

Figure 8.3(a) shows a typical contact problem of a cylindrical roller on a flat plane. Initially the contact is at a single point, and then progresses as the

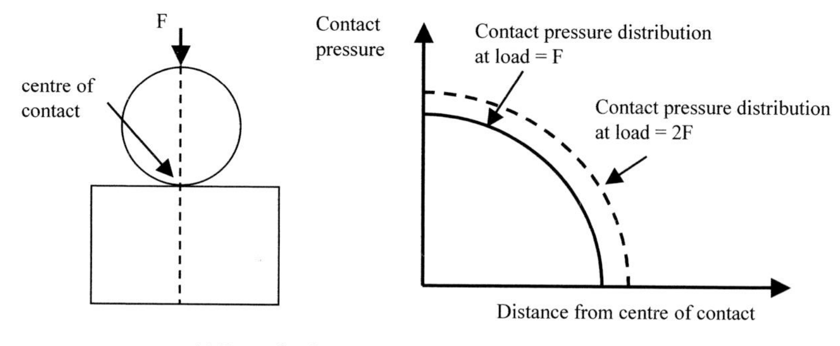

(a) Example of contact of a roller on a plane

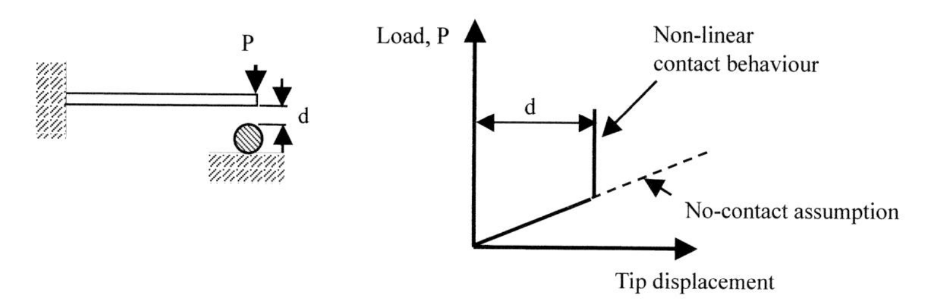

(b) Example of a cantilever in contact with a rigid support

Fig. 8.3 Examples of boundary nonlinearity problems

load is increased. The increase in the contact area and the change in the contact pressure are not linearly proportional to the applied load. Another example is shown in Fig. 8.3(b) where the tip of a cantilever beam comes into contact with a rigid surface.

In real-life engineering problems, a combination of the three nonlinearities discussed above may occur; for example, problems in which the stress–strain relationships are nonlinear as well as undergoing large displacements and strains. Examples of combined material and geometric nonlinearities include the collapse analysis of metal structures and metal forming.

8.4 NONLINEAR FE PROCEDURES

In nonlinear FE formulations, the main strategy is to break down the loading history into a series of simpler 'piecewise–linear' or weakly nonlinear steps. A combination of load incrementation and iterative procedures is used in commercial FE codes to arrive at the final solution, often automated by the code developers and requiring little or no interaction from the user. The robustness and accuracy of these solution procedures have a large influence on the reliability of the FE solutions.

In conventional FE analysis of linear elastic problems, the system of linear algebraic equations to be solved by the computer can be expressed as

$$[K][u] = [F] \tag{8.1}$$

where $[K]$ is the stiffness matrix, $[u]$ is the displacement vector, and $[F]$ is the vector of applied external forces.

The main feature of linear problems is that the full load can be applied instantaneously and the loading history is irrelevant. In other words, the displacements are linearly dependent on the loads and, if required, the solution can be easily scaled. This also means that most material properties can be scaled.

A brief description of nonlinear procedures used in FE formulations is presented below. Further details of nonlinear FE procedures can be found in many FE textbooks (**10**)–(**13**).

8.4.1 Newton–Raphson iterative method

The Newton–Raphson iterative method is a well-established effective numerical method of arriving at the solution of nonlinear equations. It works by guessing a trial solution, and then successively improving the initial guess by using the slope of the load–displacement curve. The nonlinear curve is effectively approximated by a series of suitable tangents. Figure 8.4 shows a schematic representation of the Newton–Raphson method.

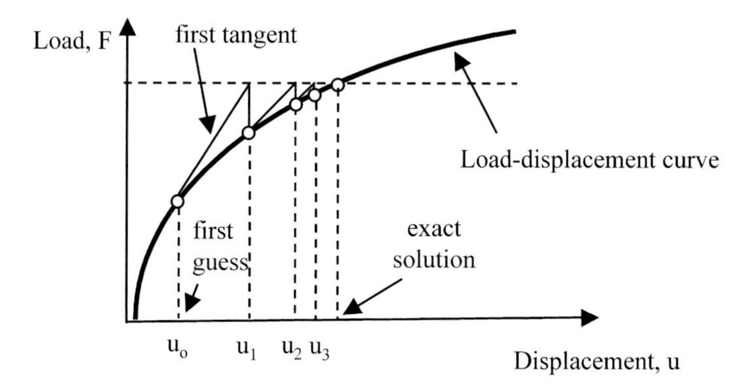

Fig. 8.4 Schematic representation of the Newton–Raphson method

Several well-established numerical techniques already exist for using iterations to solve nonlinear equations of the form

$$f(x) = 0 \tag{8.2}$$

where the function $f(x)$ is a nonlinear function of x. In the Newton–Raphson (also known as Newton) method a trial solution, x_i, is first used, which is reasonably close to the exact solution. The next trial solution, x_{i+1}, is then estimated using the slope $(\mathrm{d}y/\mathrm{d}x)$ of the curve at the point x_i, such that

$$x_{i+1} = x_i - \frac{f(x_i)}{(\mathrm{d}y/\mathrm{d}x)_i} \tag{8.3}$$

Iterations are performed until the exact solution can be found to a specified degree of accuracy or tolerance, that is, when the right-hand side of the nonlinear equation is very close to zero. Note that this approach only works if the nonlinear function $f(x)$ is differentiable. The nonlinear curve is effectively approximated by a series of suitable tangents. Experience has shown that the Newton–Raphson method can be successfully applied to nonlinear FE analysis, provided that two conditions are met:

(a) The initial guess is not very far from the exact solution; and
(b) The slope of the nonlinear load–displacement curve does not change its sign.

If the above conditions are not met, the solution may not converge. A typical example of how nonconvergence can easily occur is shown in Fig. 8.5 where the nonlinear curve $f(x)$ changes slope, and the first guess of x, that is, x_0, produces a worse approximation for x_1. This demonstrates that the Newton

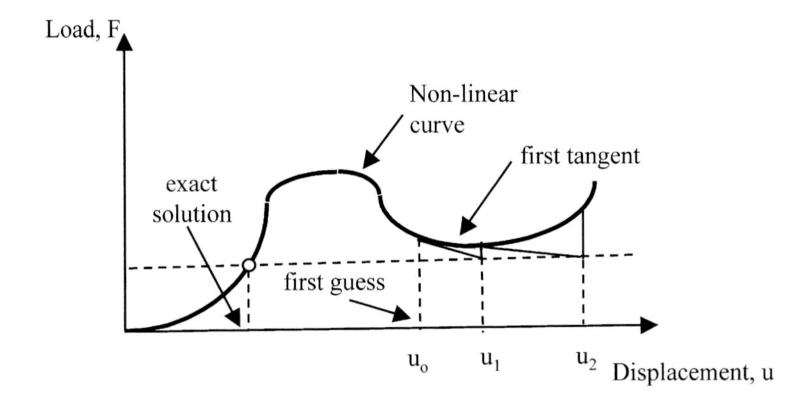

Fig. 8.5 Possible nonconvergence of the Newton–Raphson method

method can easily fail in situations where there is a minimum or a maximum in the curve, and the first guess is not sufficiently close to the exact solution. More sophisticated methods are needed for such problems.

Because the Newton–Raphson method requires the calculation of the slope of the load–displacement curve, that is, the stiffness matrix $[K]$, it can become expensive in terms of computation time if the slope is updated after each iteration. However, if the slope is kept constant during the iterations, a slower convergence usually occurs, but at a lower cost per iteration. In practice, the best compromise is to keep the slope constant for a number of iterations before updating it. Keeping the slope constant for successive iterations is usually referred to as the 'Modified Newton–Raphson' method.

Example

As an example of how the Newton–Raphson method works, consider the solution to the following nonlinear equation.

$$x^3 = c \tag{8.4}$$

The nonlinear function can be written as

$$f(x) = x^3 - c = 0 \tag{8.5}$$

and the slope, that is, the differential of $f(x)$, is given by

$$\frac{\mathrm{d}f(x)}{\mathrm{d}x} = 3x^2 \tag{8.6}$$

Using the Newton method for $c = 12.0$, and starting from an initial guess of $x_0 = 1.0$, the first approximation is obtained as

$$x_1 = 1.0 - \frac{(1.0)^3 - 12.0}{3(1.0)^2} = 4.6667 \tag{8.7}$$

The second approximation, x_2, is similarly obtained from x_1, as

$$x_2 = 4.6667 - \frac{(4.6667)^3 - 12.0}{3(4.6667)^2} = 3.2948 \tag{8.8}$$

If this procedure is continued, the following approximations will be obtained.

$$\begin{aligned}
x_3 &= 2.5650 \\
x_4 &= 2.3179 \\
x_5 &= 2.2898 \\
x_6 &= 2.2894 \text{ (which is the exact solution)}
\end{aligned}$$

Therefore, convergence has been achieved after six iterations. Note that a quicker convergence would be obtained if the initial guess of the value x, that is, x_0, had been closer to the exact solution. Although this problem is mathematically simple, it demonstrates how the Newton method can be applied to solve nonlinear equations.

8.4.2 Load incrementation procedure

In this procedure, the total applied load is divided into small increments and each increment is applied individually. Provided the increments are small, the material behaviour may be assumed to be linear during the load increment, and a new stiffness matrix $[K]$ can be used in each increment. The loading history is therefore treated as 'piecewise–linear'. Figure 8.6 shows a schematic representation of the incremental procedure, which often tends to drift away from the equilibrium path.

In order to satisfy equilibrium, iterations should be performed within each load increment to ensure that the equilibrium is satisfied, that is, any out-of-balance forces must remain below a specified tolerance. The procedure terminates when the final load is reached and equilibrium is satisfied. The total deformation of the body is then calculated as the sum of the deformations associated with each load increment.

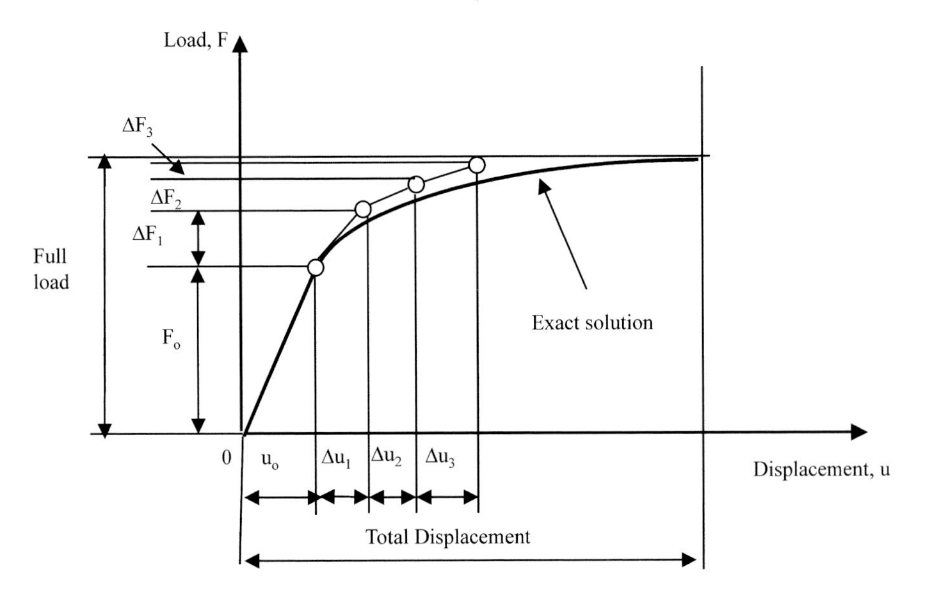

Fig. 8.6 Schematic representation of the load incrementation procedure

The incremental procedure can be summarized in the following steps:

1. Start from an 'initial state' (e.g. the elastic solution), using initial external loads $[F_0]$ and initial displacements $[u_0]$.
2. Divide the total external load into M small increments such that

$$[F_{\text{total}}] = [F_0] + \sum_{i=1}^{M} [\Delta F_i] \tag{8.9}$$

   Note that the size of the load increment does not have to be constant, and may be automatically increased or decreased by the FE program based on an assessment of the changes between successive increments.
3. Apply just one load increment $[\Delta F_i]$ and solve the equations to obtain the resulting displacement increment $[\Delta u_i]$ such that

$$[\Delta u_i] = [K_{i-1}]^{-1}[\Delta F_i] \tag{8.10}$$

4. Perform iterations to check that equilibrium is satisfied.
5. Apply the next load increment. Calculate a new updated stiffness matrix $[K_i]$ for each new load increment, or alternatively use the old value of $[K]$.
6. Repeat the iterative procedure for each load increment.
7. Stop when the full applied load is reached.
8. Calculate the total (final) displacements by adding all the displacement increments as follows.

$$[u_{\text{total}}] = [u_0] + \sum_{i=1}^{M} [\Delta u_i] \tag{8.11}$$

The stiffness matrix calculated for each increment is often referred to as the 'tangent stiffness matrix', that is, the tangent of the load–displacement curve. The incremental procedure is similar to the Euler method and the Runge–Kutta technique used for the integration of a system of linear or nonlinear differential equations.

8.4.3 Iterative procedure

It is important to perform a check, within each load increment, to ensure that the solution is acceptable, that is, equilibrium is satisfied, before proceeding to the next load increment. This is usually done by computing the residual or out-of-balance forces within the structure, and reducing them to a negligible value. Figure 8.7 shows a schematic representation of the iterative procedure

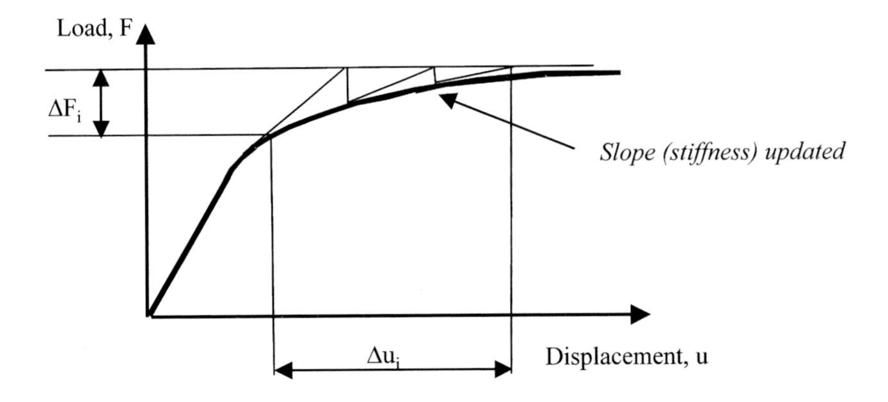

(a) Slope updated after each iteration

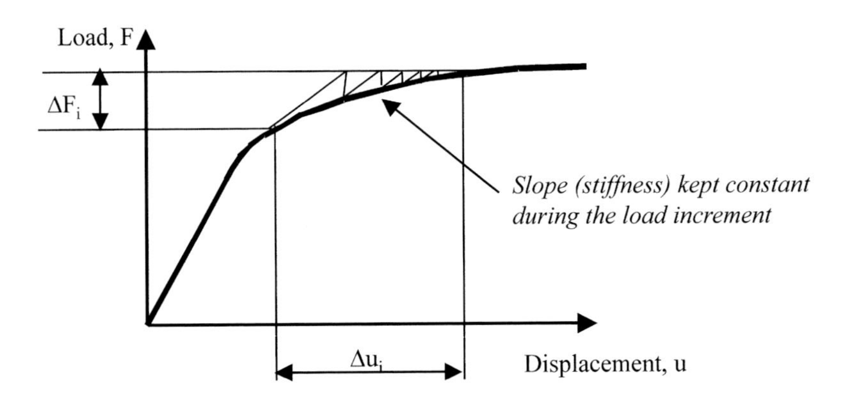

(b) Slope kept constant during the iterations

Fig. 8.7 Schematic representation of the iterative procedure

in which the slope of the curve is either updated after each iteration or kept constant. In some problems, keeping the slope constant may require more iterations for convergence, but can be computationally less expensive than updating the slope.

An outline of the iterative procedure to be performed in each load increment is presented below:

1. Using the strain–displacement and stress–strain relationships, calculate the internal forces, $[\Delta F_{\text{internal}}]$, resisting the load increment.

2. Determine the residual force vector, $[\Delta F_{\text{residual}}]$, as the difference between the applied external load increment, $[\Delta F_{\text{external}}]$, and the resisting internal force, $[\Delta F_{\text{internal}}]$, that is,

$$[\Delta F_{\text{residual}}] = [\Delta F_{\text{external}}] - [\Delta F_{\text{internal}}] \tag{8.12}$$

3. Check convergence.
   (a) *Convergence*: Convergence is achieved if the nodal values of $[\Delta F_{\text{residual}}]$ are negligible, that is, below a certain tolerance, defined either by the user or automatically by the FE code. Typically, the 'norm' of the vector is used to check convergence, which is defined as the sum of the square roots of the nodal values, as

$$\text{Norm}\ [\Delta F_{\text{residual}}] = \sqrt{\left\{ \sum_{i=1}^{N} \left(\left[\Delta F_{\text{residual}} \right]_i \right)^2 \right\}} \tag{8.13}$$

   where subscript i represents the nodal value and N is the total number of nodes.
   (b) *No convergence*: If convergence is not achieved, that is, the ratio $[\Delta F_{\text{residual}}]$ to $[\Delta F_{\text{external}}]$ is not smaller than a specified tolerance, then a correction to the displacement vector is necessary. The residual force $[\Delta F_{\text{residual}}]$ can be used to obtain a correction to the displacement, as

$$[\Delta u_{\text{correction}}] = [K^{-1}][\Delta F_{\text{residual}}] \tag{8.14}$$

   Using this correction, obtain a new (improved) value of the displacement vector and go back to step (3).
   Repeat this correction procedure until the displacement corrections or the residual force vector $[\Delta F_{\text{residual}}]$ become negligible, that is, below a given tolerance.
4. Store all the values of displacements and forces at all the nodal points and proceed to the next load increment.

Several procedures exist for accelerating the rate of convergence. Standard incremental–iterative procedures require that the initial trial solution is sufficiently close to the exact solution, and the load–deformation slope remains of the same sign during the loading path. In situations where the slope may change sign, for example, in a material softening or a snap-through behaviour, more sophisticated numerical algorithms should be employed (**13**, **14**).

8.4.4 Displacement control

In the load incrementation procedure, the solution is driven by 'load control', that is, the structure is only allowed to deform by a single load increment at a time. This does not work when the load–displacement tangent becomes horizontal, because the load must decrease in order to satisfy equilibrium. In such problems, 'displacement control' should be used where the displacement of a specified node is limited to a small displacement. This enables the FE solution to follow the load path correctly in such problems. In some problems, such as those associated with geometric nonlinearity, snap-back, and snap-through problems, a mixture of load control and displacement control may be the best approach.

8.5 Issues Related to Nonlinear Analysis

Before attempting the solution of nonlinear problems, the user must carefully consider a number of issues related to nonlinear analysis. The following questions should be addressed by the user when analysing nonlinear problems:

- *Material properties*: How accurate are the material properties? Would the solution be sensitive to the accuracy of the material data? If so, is it worth performing experimental tests to obtain a reliable set of material data?
- *Initial conditions*: Is there an initial state of stress/strain, for example, locked-in residual stresses? If so, how can they be accounted for?
- *Uniqueness of solution*: Is there a possibility of more than one solution existing, for example, an unstable path in buckling problems or more than one buckling mode? If so, will it be necessary to introduce some kind of an artificial imperfection to arrive at only one stable solution?
- *Load history*: Will the structural behaviour be influenced by the load history? Will the material behaviour be the same in loading and unloading?
- *Load direction*: In large displacement or large rotation problems, will the direction of the applied loads remain in the same initial direction, or will they follow the deformed shape?
- *Nonlinear assumptions*: Are the assumptions used in the nonlinear model adequate for the anticipated accuracy of the solution? Should a larger factor of safety be used to compensate for any inaccuracies or uncertainties in the solutions?

8.6 Difficulties in Modelling Nonlinear Problems

Modern FE codes are capable of handling most nonlinear problems. However, the user must be able to identify 'difficult' nonlinear problems, that is, problems where the FE code may struggle with or encounter convergence problems. Most FE codes are optimized to achieve the most accurate solution with the least number of load increments and iterations. This is usually achieved by automating the load incrementation procedure and the iterative process, so that the user can obtain a converged solution with sufficiently small tolerances, without having to interfere with the automatic process.

8.6.1 Difficulties experienced by the user

Whether using FE codes, or any type of numerical approach, there are many difficulties associated with nonlinear problems. For example,

- *Material properties*: In problems involving nonlinear material behaviour of real-life structures, accurate material properties are often difficult to obtain. It may even become necessary, before proceeding with the nonlinear analysis, to commission a series of experimental tests to obtain a reliable set of accurate material data. The FE analyst must be able to interpret the data from the experimental tests and correctly feed them into the FE code.
- *Experience of the analyst*: Most analyses of real-life nonlinear problems heavily rely on experience and past knowledge. This makes it difficult for an inexperienced analyst to attempt to analyse nonlinear problems for the first time. Furthermore, experience in one type of nonlinearity, for example, plastic behaviour of metals, does not necessarily throw light on a different type of nonlinearity, such as buckling.
- *Direction of load*: In nonlinear problems where the displacements can be large, a decision has to be made by the user whether to keep the applied loads in the same initial direction (conservative loads), or to allow them to follow the deformed shape (follower loads). The loading type may significantly affect the solution accuracy, and in some problems it may be difficult to establish the direction of the follower load.
- *Accuracy of solutions*: In the absence of other alternative solutions, it is difficult to judge the accuracy of the final solutions.

8.6.2 Difficulties encountered by the FE code

To appreciate the type of problems that may be encountered by the FE code in solving difficult nonlinear problems, the following observations can be made:

- *Increment size*: It is difficult for the FE code to decide in advance, for all nonlinear problems, what size of load increment and how many increments should be used. Using unnecessarily small increment sizes can significantly increase the computation time. FE codes often automate the load incrementation procedure, starting from a very small increment, for example, 1 percent of the applied load, and then continually adjusting it as required.

- *Convergence criteria*: The default automatic convergence criterion used in the FE code may not be suitable for all types of nonlinear problems. The user should be aware of the convergence criteria in the FE code, and, if necessary, be able to change them for a given problem. However, this is only advisable for experienced users. For example, in a problem with very high localized stresses, such as around a crack tip region, the calculation of the norm of the residual force vector for the whole structure may not be a good representation of the local stress changes in the high stress region (the crack region), particularly if the crack region is only a small part of the overall geometry.
- *Changing sign of the load–displacement slope*: If the slope of the nonlinear curve changes its sign during the loading, for example, in a work softening situation or in a snap-through problem, the standard incremental–iterative procedures may fail. More complex routines have to be introduced in such problems.
- *Non-unique solutions*: In some problems, particularly where geometric nonlinearity is involved, the converged solutions may not be unique, that is, there may be more than one solution that satisfies equilibrium. Buckling behaviour is an example of such problems. For example, a strut subjected to an axial load may follow three different load paths: simple compression, buckling to the left, or buckling to the right.
- *Complex user input*: In complex nonlinear material behaviour, it may become necessary to specify the material law through 'subroutines' or more complex input coding. Because this means that the user has to delve deeper into the FE code manuals, the FE code must have a clear set of documentation and allow the user flexibility in specifying nonstandard material laws, without compromising the efficiency of the automated solution procedures.

Therefore, more caution should be exercised in the solution of nonlinear problems using the FE technique. The user must always apply engineering judgement to decide whether the solution is physically acceptable. In practice, when faced with a new nonlinear problem, the user should perform convergence studies, for example, investigate the effect of changing the mesh refinement, changing the sizes of the load increment, or using different tolerances.

Owing to the possible difficulties that may be encountered when analysing nonlinear problems, it is risky for an inexperienced FE user to tackle a nonlinear application without at least running a few nonlinear benchmarks, with a known solution, relevant to the type of nonlinearity encountered in the application. A benchmark can also be used to illustrate the use of the FE code to ensure that the code is used correctly.

8.7 SUMMARY OF KEY POINTS

- The FE strategy in obtaining solutions to nonlinear problems is to reduce the nonlinear loading history of the structure to a sequence of linear or weakly nonlinear increments.
- The applied load is usually applied in small load increments starting from an initial value and an iterative procedure is used to ensure that equilibrium is satisfied at each load increment.
- In nonlinear applications, more care must be taken to use accurate material properties and a consistent set of units for all variables, because the solutions cannot be linearly scaled.
- There are difficulties that may be encountered in analysing and modelling nonlinear problems using the FE technique. The user should use engineering judgement to check that the FE solutions are reliable.
- In practice, when faced with a nonlinear problem for the first time, the user should run a nonlinear benchmark test and perform convergence studies, for example, investigate the effect of changing the load increment size or mesh refinement.

CHAPTER 9

Thermal Problems

9.1 INTRODUCTION

This chapter demonstrates how the variational formulation in FE analysis can be used for nonstructural problems such as heat conduction problems. An outline of the numerical schemes used for time-marching in time-dependent problems is presented.

Thermal analysis using FE analysis can be used to calculate the temperatures in the structure and then feed them into the structural analysis to calculate thermal stresses and strains. Thermal strain is defined as

$$\varepsilon_{\text{thermal}} = \alpha(\Delta T) \tag{9.1}$$

where $\varepsilon_{\text{thermal}}$ is the thermal strain, α is the coefficient of thermal expansion ($^\circ C^{-1}$), and ΔT is the change in temperature from a datum reference temperature ($^\circ C$). Note that the thermal strain can usually be superimposed on the mechanical strain.

9.2 MODES OF HEAT TRANSFER

9.2.1 Conduction

This mode of heat transfer occurs as heat flows through a solid body from a high-temperature region to a low-temperature one. The rate of heat flow by conduction through a solid is proportional to the normal gradient of the temperature, such that

$$q = -kA\frac{dT}{dn} \tag{9.2}$$

where q is the rate of heat flow (W), k is the thermal conductivity, which is a material property of the solid ($W\,m^{-1}K^{-1}$), A is the cross-sectional area through which the heat flows (m^2), and dT/dn is the temperature gradient in the normal direction, n, that is, perpendicular to the surface ($K\,m^{-1}$).

9.2.2 Convection

Convection heat flow occurs as the thermal energy is transferred from a solid to a surrounding liquid or gas. Convection is either forced or natural. Forced convection occurs when the fluid is forced to flow around the solid body, whereas in natural convection, the fluid flow is caused by density variations arising from the heat transfer. The rate of heat flow, q, by convection is given by

$$q = hA(T_{\text{solid}} - T_{\text{liquid}}) \tag{9.3}$$

where h is called the heat transfer coefficient, which can be obtained by experimental testing involving the solid and the liquid (W m^{-2} K^{-1}), A is the cross-sectional area through which the heat flows (m^2), T_{solid} is the temperature at the surface of the solid body (K), and T_{liquid} is the ambient or remote temperature of the liquid (K).

9.2.3 Radiation

Radiation heat flow is the process by which thermal energy is transmitted by electromagnetic radiation, and can occur with or without the presence of a liquid or gas. The rate of heat flow, q, by radiation is given by the 'Stefan–Boltzmann law', as follows (**16**)

$$q = \varepsilon_{\text{SB}} \sigma A[(T_{\text{solid}})^4 - (T_\infty)^4] \tag{9.4}$$

where ε_{SB} is the Stefan–Boltzmann constant (W m^{-2} K^{-4}), σ is the emissivity of the surface, which is dependent on the appearance of the surface (emissivity is defined as the ratio of the heat emitted by the surface to that emitted by a perfectly black body at the same temperature, and is affected by the temperature of the surface), T_{solid} is the temperature at the boundary of the solid body (K), and T_∞ is the ambient temperature (K). Note that the FE analysis of thermal radiation requires a nonlinear analysis.

9.3 Partial differential equation for heat conduction

By considering the energy balance for a three-dimensional small element of the material with dimensions dx, dy, and dz, the partial differential equation (PDE) for heat conduction can be derived as

$$\frac{\partial}{\partial x}\left(k_x \frac{\partial T}{\partial x}\right) + \frac{\partial}{\partial y}\left(k_y \frac{\partial T}{\partial y}\right) + \frac{\partial}{\partial z}\left(k_z \frac{\partial T}{\partial z}\right) + q = \rho c_p \frac{\mathrm{d}T}{\mathrm{d}t} \tag{9.5}$$

where k_x, k_y, and k_z are the thermal conductivities in the Cartesian axes x, y, and z (W m^{-1} K^{-1}), t is the time, q is the rate of heat flow (W), ρ is the density of the solid (kg m^{-3}), and c_p is the specific heat of the solid (J kg^{-1} K^{-1}).

For simplicity, assume that heat conductivity is the same in all directions, that is, $k_x = k_y = k_z = k$. This results in the following governing partial differential equation.

$$k\left(\frac{\partial^2 T}{\partial x^2}+\frac{\partial^2 T}{\partial y^2}+\frac{\partial^2 T}{\partial z^2}\right)+q=\rho c_p\frac{\mathrm{d}T}{\mathrm{d}t} \tag{9.6}$$

For steady-state heat conduction, that is, where there is no change in temperature with time, the above equation reduces to the following PDE.

$$k\left(\frac{\partial^2 T}{\partial x^2}+\frac{\partial^2 T}{\partial y^2}+\frac{\partial^2 T}{\partial z^2}\right)+q=0 \tag{9.7}$$

9.4 Variational formulation for heat conduction

The variational formulation for thermal problems can be written in terms of a functional, F (equivalent to the *TPE* in structural problems), to be minimized with respect to temperature, as

$$F=\frac{1}{2}\int_V kA\left[\left(\frac{\mathrm{d}T}{\mathrm{d}x}\right)^2+\left(\frac{\mathrm{d}T}{\mathrm{d}y}\right)^2+\left(\frac{\mathrm{d}T}{\mathrm{d}z}\right)^2\right]\mathrm{d}V+\int_{S_1} T_i Q_i\,\mathrm{d}S_1$$
$$+\int_{S_2} h(T-T_\infty)\,\mathrm{d}S_2 \tag{9.8}$$

where V is the volume of the solid, S_1 is the surface over which the temperature or heat flux is prescribed, S_2 is the surface over which heat convection occurs, and Q_i is the heat flux at a node i with temperature T_i (W m^{-2}).

9.5 FE formulation for heat transfer problems

The independent variable (i.e. the variable to be solved by the FE formulation) in thermal problems is temperature (equivalent to displacement in structural problems). The heat flux is equivalent to the external force in structural problems.

The final system of equations to be solved by the FE solver can be written as

$$[K][T]=[Q] \tag{9.9}$$

where $[K]$ is the 'conductance' matrix (equivalent to the stiffness matrix in structural problems), $[T]$ is the vector of nodal temperatures, and $[Q]$ is the vector of nodal heat flux.

In thermal problems the following boundary conditions are used:

(a) Prescribed temperatures at points on the surface of the solid (called the 'essential' boundary conditions).
(b) Prescribed values of heat flux at the surface (called 'natural' boundary conditions). The default value of heat flux is zero, that is, perfectly insulated condition. This means that nodes without a specified heat flux will be assigned a zero heat flux value (similar to the external force vector being zero on free surfaces in structural problems).
(c) Convection of solid–liquid interface (called 'mixed' boundary conditions). This means that neither the prescribed temperature nor the heat flux are prescribed at the node. Instead a linear relationship between the temperature and flux is prescribed (similar to a spring attached to a node in structural problems).
(d) In transient heat conduction, 'initial conditions' are needed, that is, the initial temperatures at every node at the start of the time step.

9.6 One-dimensional linear heat conduction element

Consider a one-dimensional heat conduction in a bar or fin. For simplicity, linear elements with one node at either end are used, as shown in Fig. 9.1.

9.6.1 Ritz approach

Following the steps of the Ritz solution, a trial solution for the temperature variation across an element can be assumed as

$$T(x) = \left(1 - \frac{x}{L_e}\right)T_1 + \left(\frac{x}{L_e}\right)T_2 \tag{9.10}$$

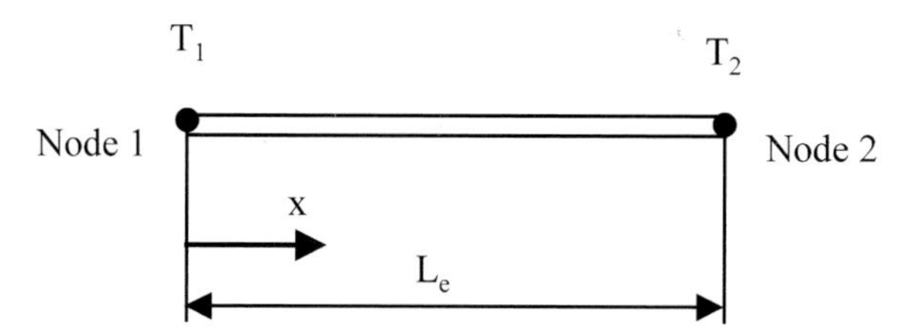

Fig. 9.1 One-dimensional heat conduction element

This trial solution satisfies the two essential boundary conditions:

(a) at node 1 (where $x = 0$), $T = T_1$;
(b) at node 2 (where $x = L_e$), $T = T_2$.

The derivative of the temperature can be obtained by differentiating the temperature function with respect to x, such that

$$\frac{\mathrm{d}T}{\mathrm{d}x} = -\frac{1}{L_e}T_1 + \frac{1}{L_e}T_2 = \frac{T_2 - T_1}{L_e} \tag{9.11}$$

The functional (equivalent to *TPE*) used for heat transfer problem, shown in equation (9.8), can be written for one-dimensional problems, as

$$F = \int_0^{L_e} \left[\frac{kA}{2} \left(\frac{\mathrm{d}T}{\mathrm{d}x} \right)^2 - Tq \right] \mathrm{d}x \tag{9.12}$$

Substituting the temperature function in the expression for F results in

$$F = \int_0^{L_e} \frac{kA}{2} \left(\frac{T_2 - T_1}{L_e} \right)^2 \mathrm{d}x - \int_0^{L_e} \left[\left(1 - \frac{x}{L_e} \right) T_1 + \left(\frac{x}{L_e} \right) T_2 \right] q \, \mathrm{d}x \tag{9.13}$$

Minimizing the functional F with respect to the temperatures, T_1 and T_2, results in

$$\begin{aligned} \frac{\partial F}{\partial T_1} &= 0 = \int_0^{L_e} \frac{kA}{2L_e^2} (2T_1 - 2T_2) \, \mathrm{d}x - \int_0^{L_e} \left(1 - \frac{x}{L_e} \right) q \, \mathrm{d}x \\ \frac{\partial F}{\partial T_2} &= 0 = \int_0^{L_e} \frac{kA}{2L_e^2} (2T_2 - 2T_1) \, \mathrm{d}x - \int_0^{L_e} \left(\frac{x}{L_e} \right) q \, \mathrm{d}x \end{aligned} \tag{9.14}$$

The above two equations can be expressed in matrix form, as

$$[k_e][T_e] = [Q_e] \tag{9.15}$$

where the element conductance (stiffness), temperature, and heat flux are defined as

$$[k_e] = \int_0^{L} \left(\frac{kA}{L_e^2} \right) \begin{bmatrix} 1 & -1 \\ -1 & 1 \end{bmatrix} \mathrm{d}x; \quad [T_e] = \begin{bmatrix} T_1 \\ T_2 \end{bmatrix}; \quad [Q_e] = \begin{bmatrix} Q_1 \\ Q_2 \end{bmatrix} \tag{9.16}$$

By adding the contributions of all the elements, the final system of equations can be written in terms of global matrices as

$$[K]_{\text{global}}[T]_{\text{global}} = [Q]_{\text{global}} \tag{9.17}$$

The stiffness matrix $[K]$ here is often referred to as the 'conductance' matrix in heat conduction problems.

9.6.2 Galerkin weighted residual approach

Alternatively, the Galerkin weighted residual method can be used directly on the partial differential equation, rather than minimizing the functional, F. The PDE for one-dimensional steady-state heat conduction is

$$kA\frac{\mathrm{d}^2T}{\mathrm{d}x^2} + q(x) = 0 \tag{9.18}$$

As in the Ritz method, the trial solution can be written as two terms as given in equation (9.10). For heat transfer problems, the Galerkin weighted residual approach requires that the PDE of equation (9.7) is multiplied by the trial functions such that

$$\int_0^{L_e} \left(kA\frac{\mathrm{d}^2T}{\mathrm{d}x^2} + q(x) \right) \mathit{Trial\ solution}(x)\,\mathrm{d}x = 0 \tag{9.19}$$

After some algebraic manipulation, an expression for each element can be derived, which is identical to the expression obtained using the Ritz method.

9.7 Transient (unsteady) heat conduction problems

In transient (i.e. time-dependent) heat conduction problems, the temperature changes with time, whereas in steady-state heat conduction, the variation of temperature with time is ignored. The final system of equations to be solved by the FE solver has to include an extra term to account for time-dependency, such that

$$[C]\left[\frac{\mathrm{d}T}{\mathrm{d}t}\right] + [K][T] = [Q] \tag{9.20}$$

where $[C]$ is the 'capacitance matrix', which contains the product of the density ρ and the specific heat c_p, as shown in the PDE of equation (9.5).

In transient heat conduction (or in any other time-dependent behaviour), it is necessary to divide the time into small time steps and 'march' the solution forward in time. There are many established time-marching numerical methods that can be used. Two popular methods used in FE formulations are called the 'explicit' and 'implicit' methods.

9.7.1 Explicit time-marching scheme (forward-difference Euler method)

This method uses a simple finite-difference approximation for the differential of the temperature with respect to time.

$$\left[\frac{dT}{dt}\right]_t = \frac{[T]_{t+\Delta t} - [T]_t}{\Delta t} \tag{9.21}$$

Substituting this expression into equation (9.20) results in:

$$[C]\left(\frac{[T]_{t+\Delta t} - [T]_t}{\Delta t}\right) + [K][T]_t = [Q]_t \tag{9.22}$$

which can be rearranged to give the following expression (called 'recurrence formula').

$$\frac{1}{\Delta t}[C][T]_{t+\Delta t} = [Q]_t + \left(\frac{1}{\Delta t}[C] - [K]\right)[T]_t \tag{9.23}$$

This scheme is known as 'forward-difference' because it arrives at the new value of the temperature, $T_{t+\Delta t}$, by moving forward along the tangent at the previous point T_t, as shown schematically in Fig. 9.2(a). This scheme is also called 'explicit' because the new values of the temperature are immediately calculated from the old values (the initial values), that is, the new prediction of temperature, $T_{t+\Delta t}$, does not appear in the right-hand side of the equations.

The main advantage of the explicit Euler method is that it is simple to program and computationally economical because the conductance (stiffness) matrix does not appear on the left-hand side. This allows the new temperature to be calculated without having to solve a new system of equations. However, the explicit method suffers from the disadvantage that it needs very small time steps to remain stable, that is, if the time step is large, instability may occur. Explicit schemes are called 'conditionally stable', that is, they are only stable below a critical length of time step, usually given as

$$(\Delta t)_{\text{critical}} \leq \frac{2\rho c_p}{D\pi^2 k} L^2 \tag{9.24}$$

where D is either 1, 2, or 3 corresponding to one-dimensional, two-dimensional, or three-dimensional problems, respectively, and L is 'characteristic length', usually determined as the minimum distance between any two nodes in the FE mesh.

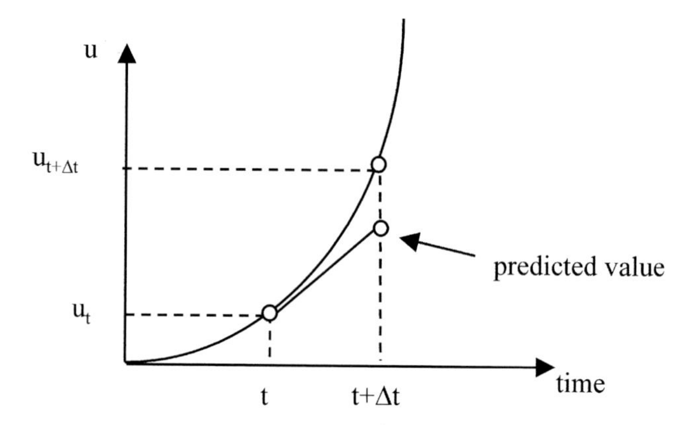

(a) Forward-difference (explicit) scheme

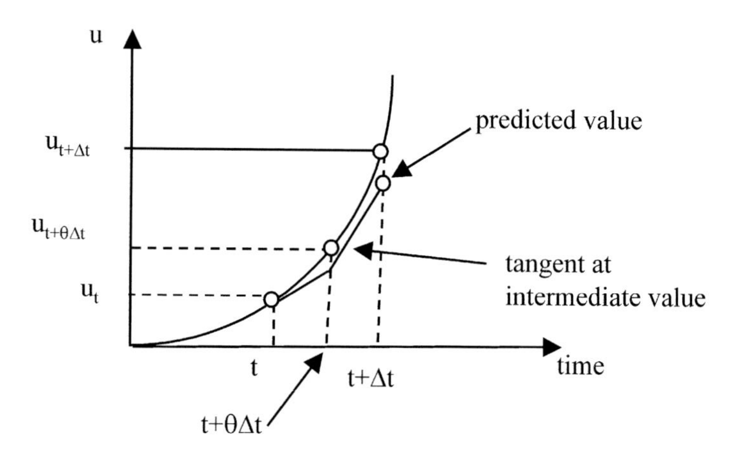

(b) Backward-difference scheme

Fig. 9.2 Time-marching schemes

9.7.2 Implicit/semi-implicit time-marching scheme (backward-difference theta method)

This is a more general approach where the slope at the initial point, T_t, and the slope at an intermediate point, $T_{t+\theta\Delta t}$, are used to obtain the solution for the temperature at the end of the time step, $T_{t+\Delta t}$, as

$$\frac{[T]_{t+\Delta t} - [T]_t}{\Delta t} = f([T]_{t+\theta\Delta t}, t + \theta\Delta t) \qquad (9.25)$$

where

$$[T]_{t+\theta\Delta t} = \theta[T]_{t+\Delta t} + (1-\theta)[T]_t \tag{9.26}$$

which can be expressed as

$$\left(\frac{1}{\Delta t}[C] + \theta[K]\right)[T]_{t+\Delta t} = (1-\theta)[Q]_t + \theta[Q]_{t+\Delta t} + \left(\frac{1}{\Delta t}[C] - (1-\theta)[K]\right)[T]_t \tag{9.27}$$

The parameter θ is effectively a 'weighted average' of the approximations at the start and end of the time interval. Figure 9.2(b) shows a schematic representation of this method where the curve is approximated at first by the straight line tangential to the curve at point T_t, which then changes to another straight line tangential to the curve at the intermediate point $T_{t+\theta\Delta t}$, until the end of the time step is reached at point $T_{t+\Delta t}$.

Choosing the value of θ has important implications for the stability of this time-marching scheme. For $\theta > 0.5$, this method becomes 'unconditionally stable' for all values of Δt, that is, it can tolerate large time steps. When $\theta = 0.5$, the theta method becomes the 'Crank–Nickolson method', which is an established numerical procedure.

When $\theta = 1$, that is, the slope of the curve at the future temperature, $T_{t+\Delta t}$, is used, the theta method becomes the 'implicit' method, which can be written as

$$\left(\frac{1}{\Delta t}[C] + [K]\right)[T]_{t+\Delta t} = [Q]_{t+\Delta t} + \left(\frac{1}{\Delta t}[C]\right)[T]_t \tag{9.28}$$

It can be shown that the implicit method is unconditionally stable. This means that this method remains stable even if large time steps are used, although small time steps are more accurate. The main disadvantage of the implicit or semi-implicit methods is that the equations have to be solved at every time step. However, because larger time steps than the explicit method can be tolerated, the solution can be obtained with relatively fewer time steps.

It should be noted that when $\theta = 0$, that is, no intermediate point is used, the theta method becomes the explicit Euler method.

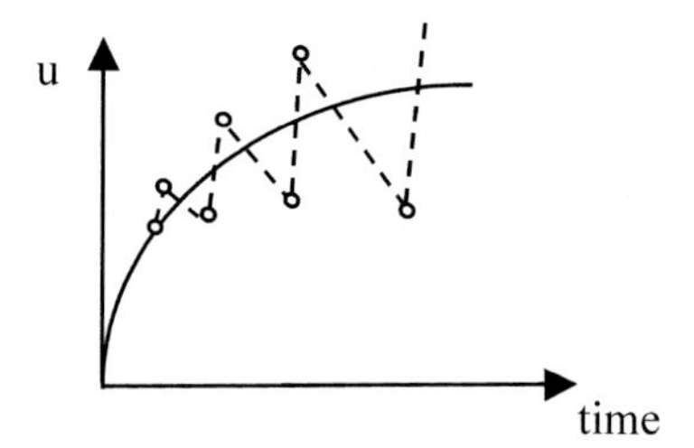

Fig. 9.3 Solution instability

9.7.3 Convergence and stability of solutions

A robust time-marching algorithm must ensure that the solution is both convergent and stable. Convergence and stability are defined as follows:

(a) *Convergence* is achieved if the numerical predictions approach the exact solution as the size of the time interval approaches zero.

(b) *Stability* is achieved if the numerical errors made at a given time interval do not cause increasingly larger errors as the solution is marched forward in time. Instead, the numerical errors should damp out with time. Figure 9.3 shows a schematic representation of instability.

It should be emphasized that the stability criteria and the convergence criteria are not necessarily related to each other. Therefore, to ensure stable and accurate solutions, the choice of the time step must be guided simultaneously by the stability criteria and the accuracy.

9.8 Summary of Key Points

- Thermal problems and other nonstructural problems can also be analysed using the FE formulation based on the Ritz method or the Galerkin weighted residual method.
- In thermal problems, temperatures replace displacements, and heat flux replaces the external forces.
- In explicit time integration schemes, the solution is based on the previous time t, whereas in implicit schemes, the solution is based on the new time $t + \Delta t$. Explicit schemes need very small time steps to remain stable, whereas implicit schemes are unconditionally stable.

CHAPTER 10

Examples of FE Applications

In Chapter 7, several practical guidelines for the application of FE analysis were discussed. In this chapter, a number of practical problems are solved using FE analysis. The examples are chosen to illustrate various aspects of FE analysis, such as choosing the correct type of element configuration, using symmetry to reduce the size of a problem, effect of mesh refinement, preventing rigid body motion, and simplifying geometric features.

All of the problems assume a linear elastic behaviour. To determine whether plasticity should be included in the analysis, the FE analysis can be performed first as an elastic analysis in order to obtain the maximum value of the equivalent (von Mises) stress. If this value exceeds the yield stress of the material, then a nonlinear plasticity analysis will be needed. It is important to note that the maximum values of the Cartesian stress components may well exceed the yield stress, but plastic behaviour is governed by the equivalent stress, σ_e, which is often based on the von Mises yield criterion given by the following equation for two-dimensional problems.

$$\sigma_e = \frac{1}{\sqrt{2}}\sqrt{(\sigma_{xx}-\sigma_{yy})^2+(\sigma_{xx}-\sigma_{zz})^2+(\sigma_{yy}-\sigma_{zz})^2+6(\sigma_{xy})^2} \quad (10.1)$$

The decision whether to include plasticity or not also depends on whether the plastic region is localized to a very small region away from the area of interest.

10.1 Example 1: Perforated plate

Problem definition

Consider a square plate of side length L and thickness t (in the z-direction) with a central circular hole of diameter D, subjected to a uniaxial stress σ_o, as shown in Fig. 10.1. The numerical values used are $L = 100$ mm, $D = 20$ mm, $t = 5$ mm, and $\sigma_o = 100$ MPa. The objective of the analysis is to determine the stress concentration around the hole.

Geometry

Because the plate thickness (in the z-direction) is small, 2D plane stress conditions are applicable. The plate (both geometry and loads) is

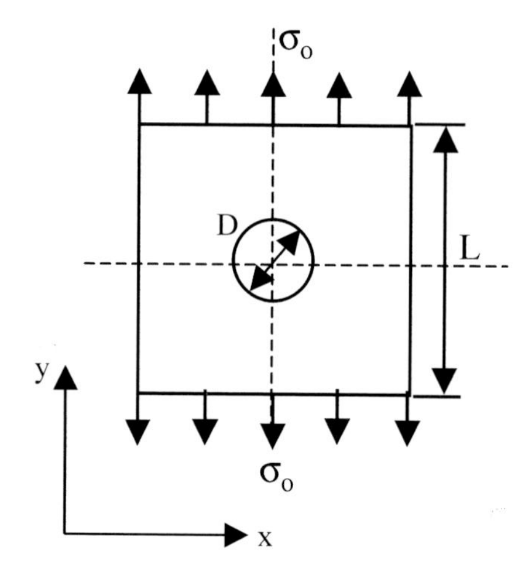

Fig. 10.1 Perforated plate example

symmetrical about the horizontal and vertical axes. Therefore, only a symmetrical quarter-model needs to be modelled, as shown in Fig. 10.2.

Material properties

Assuming an elastic analysis, the material properties needed are Young's modulus (E) and Poisson's ratio (ν). The values used here are $E = 200$ GPa and $\nu = 0.3$. If the load is high enough to cause local plasticity around the hole, the elastoplastic stress–strain curve, or at least the yield stress (σ_y), must also be specified.

Displacement boundary conditions

On the axes of horizontal and vertical symmetry, the nodes can only slide along the symmetry lines, that is, displacements perpendicular to the axes of

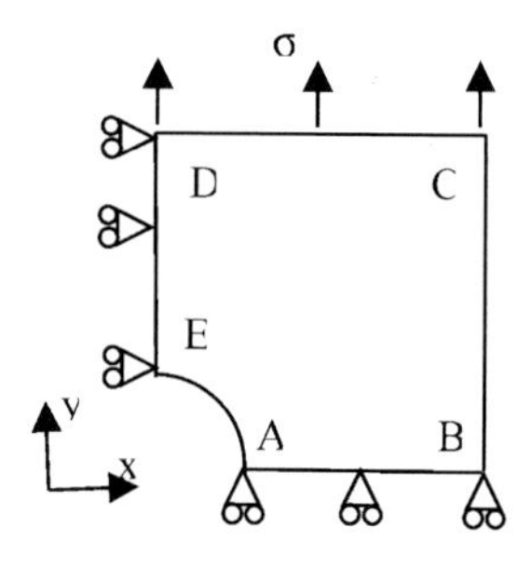

Fig. 10.2 Symmetrical quarter of the perforated plate

symmetry are prevented. Therefore, in this problem, there are two sets of displacement boundary conditions, as follows:

(a) zero y-displacements (roller conditions) specified on line AB;
(b) zero x-displacements (roller conditions) specified on line DE.

Applied loads
A uniform tensile stress (distributed load), σ_o, is specified at the top surface (line CD).

FE model
Two-dimensional plane stress isoparametric quadratic elements can be used here. Either eight-node quadrilaterals or six-node triangles, or a combination of the two, can be used. Since stress concentration is expected around the hole, mesh biasing should be specified around the hole region. If the yield stress is known, a plasticity check can be performed by checking the maximum value of the equivalent (von Mises) stress.

The analytical solution for the stress in the direction of the applied load in a perforated infinite plate is given by (**17**)

$$\sigma_{yy} = \frac{1}{2}\sigma_o\left[2.0 + \left(\frac{R}{x}\right)^2 + 3\left(\frac{R}{x}\right)^4\right] \tag{10.2}$$

where x is the distance from the centre of the hole. Note that since the plate used in this problem is not infinite, the computed stresses will be expected to be slightly higher than those predicted by the analytical solution.

To demonstrate the effect of mesh refinement on the accuracy of the FE solutions, a number of meshes are used, ranging from 2 to 32 elements, as shown in Fig. 10.3. Also, two types of elements are used: four-node linear elements and eight-node quadratic elements.

Figures 10.4 and 10.5 show comparisons of the FE and analytical solutions for various mesh densities for the four-node elements and eight-node elements (with 2×2 integration points), respectively. It can be seen that the FE solutions converge to the analytical solution as the mesh density is increased and that the quadratic eight-node elements provide better accuracy than the corresponding linear four-node elements.

The deformed shape for the 32 quadratic element mesh is shown in Fig. 10.6 where the deformations are exaggerated by multiplying them by a factor greater than 1. The deformed shape is useful in checking that the

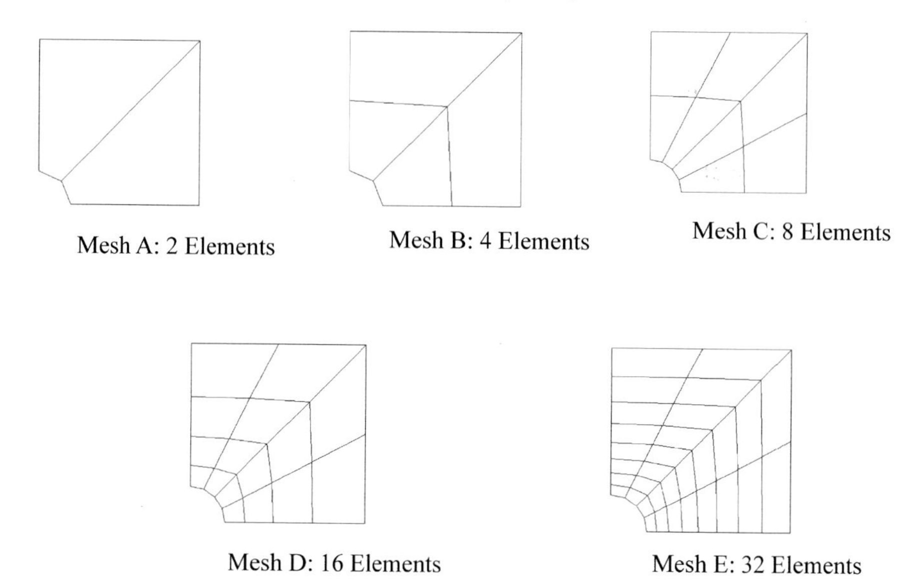

Fig. 10.3 Finite element meshes used for the perforated plate example

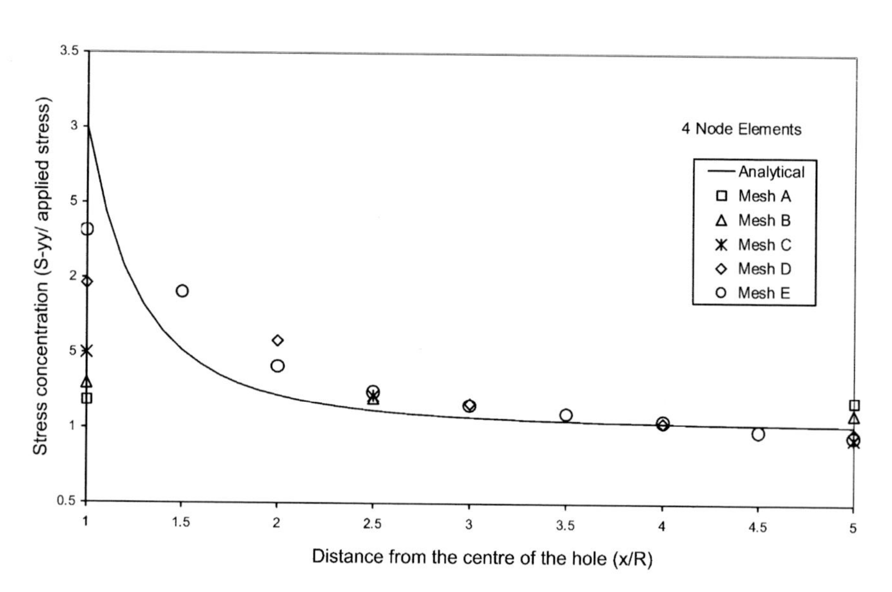

Fig. 10.4 Comparison of FE and analytical solutions for the perforated plate example (four-node elements)

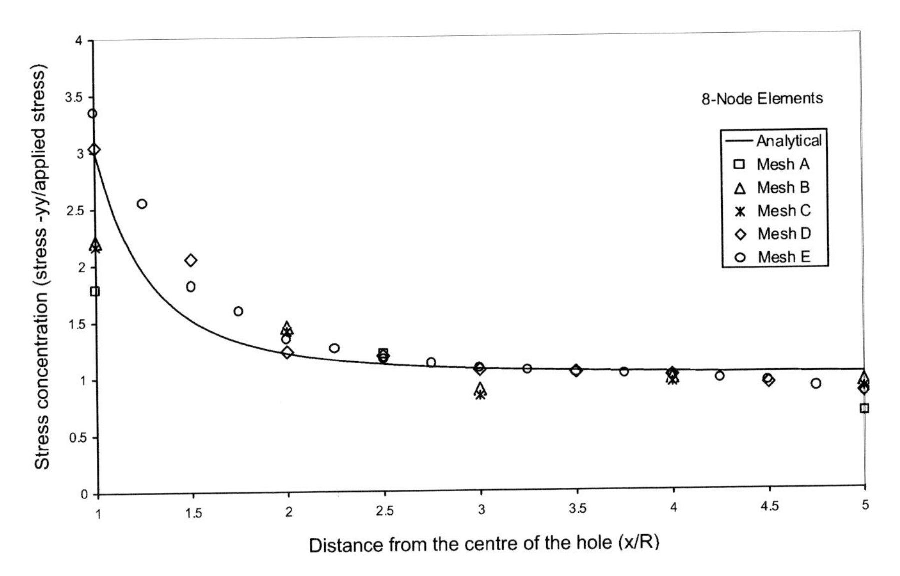

Fig. 10.5 Comparison of FE and analytical solutions for the perforated plate example (eight-node elements)

overall deformation of the body has followed the prescribed boundary conditions, that is, the left and bottom sides slide along the axes of symmetry.

The stress contour plot for the vertical stress is shown in Fig. 10.7, where, as expected, the highest stresses occur in the vicinity of the hole.

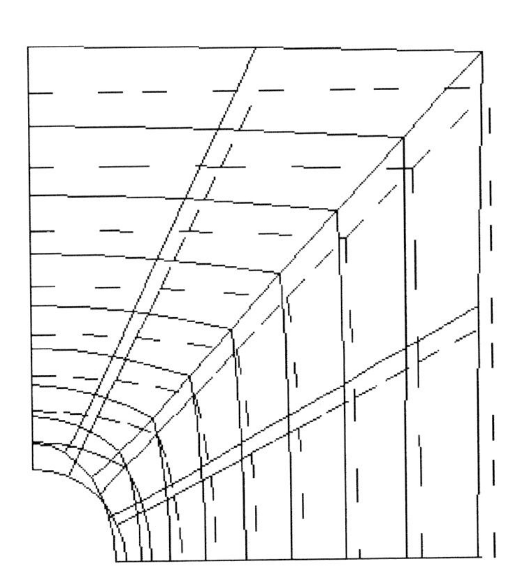

Fig. 10.6 Exaggerated deformed shape (solid lines) for the perforated plate example

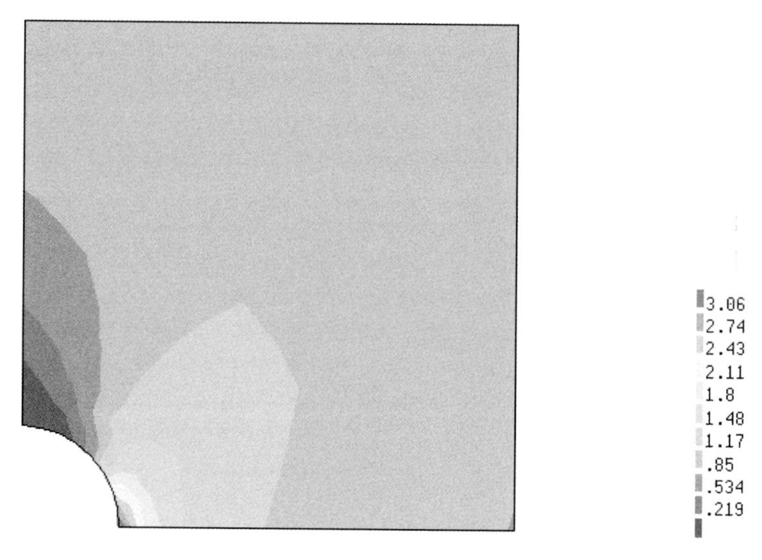

Fig. 10.7 Stress contour plot (vertical stress) for the perforated plate example
(See colour plate section.)

10.2 EXAMPLE 2: CANTILEVER BEAM

Problem definition

Consider a cantilever beam of length L built in at one end and subjected to a concentrated force F at the other end, as shown in Fig. 10.8. The beam has a square cross-sectional area of side length t. The numerical values used are: $L = 2$ m, $t = 0.1$ m, and $F = 1$ kN. The objective of the analysis is to obtain the overall deflection of the beam.

Geometry

Because there is no symmetry in this problem, the whole geometry has to be modelled. The geometry can be modelled with beam elements because the geometry and loads satisfy beam bending conditions, that is, the geometry is

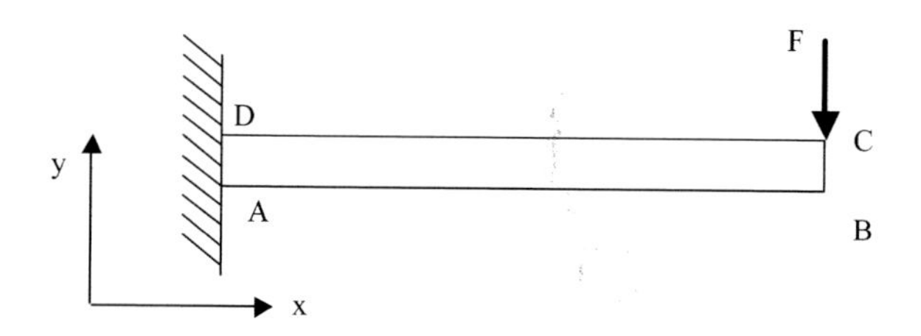

Fig. 10.8 Cantilever beam example

long, slender, and subjected to only transverse loads. However, it is also possible to model this problem with 2D plane stress elements since the thickness in the z-direction is sufficiently small.

Material properties

Assuming an elastic analysis, the material properties needed are only Young's modulus (E) and Poisson's ratio (ν). The values used here are $E = 200$ GPa and $\nu = 0.3$.

Boundary conditions

The cantilever is built-in at the left-hand side. If beam elements are used, then both the displacement and the slope at the built-in node must be prescribed as zero. If 2D plane stress elements are used, then all nodes on line AD must have zero displacements in the x- and y-directions, which automatically enforce the built-in condition. Note that slope is not a variable in 2D continuum elements.

Applied loads

A point load of magnitude F is applied to point C. If a 2D plane stress model is used, this point force can either be applied at point C, or distributed along the line BC.

FE model

As discussed above, two types of elements can be used to model this problem: beam elements or 2D plane stress elements. Of course, it is always possible to model this problem using 3D elements, but that would be unnecessary and would consume more computation time without improving the accuracy. Figure 10.9 shows a three-node beam element mesh and an alternative eight-node 2D plane stress element mesh with 2×2 integration points.

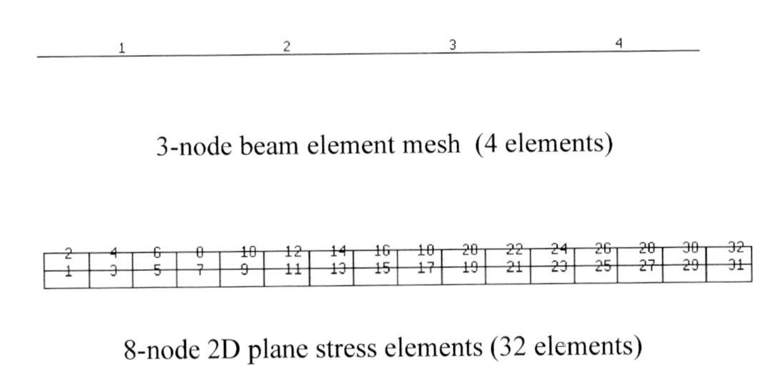

Fig. 10.9 Finite element meshes used for the cantilever beam problem

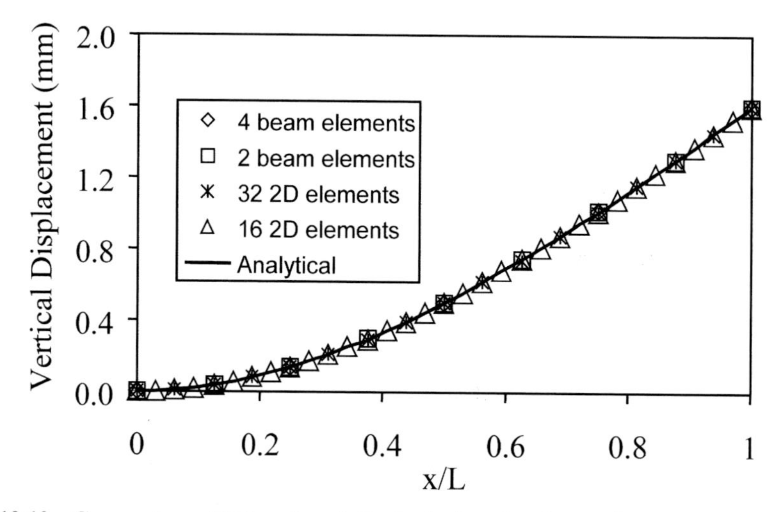

Fig. 10.10 Comparison of FE and analytical solutions for the cantilever beam problem

The analytical solution for the vertical displacement, v, in a cantilever beam can be derived from beam bending theory as (**9**)

$$v = \frac{FL^3}{EI}\left[\frac{1}{6}\left(\frac{x}{L}\right)^3 - \frac{1}{2}\left(\frac{x}{L}\right)^2\right] \tag{10.3}$$

Figure 10.10 shows a comparison of the FE and analytical solutions for a number of beam and 2D plane stress meshes, where it is clear that the FE solutions are in good agreement with the analytical solutions, even when a relatively small number of elements are used. The deformed shapes for the cantilever are shown in Fig. 10.11.

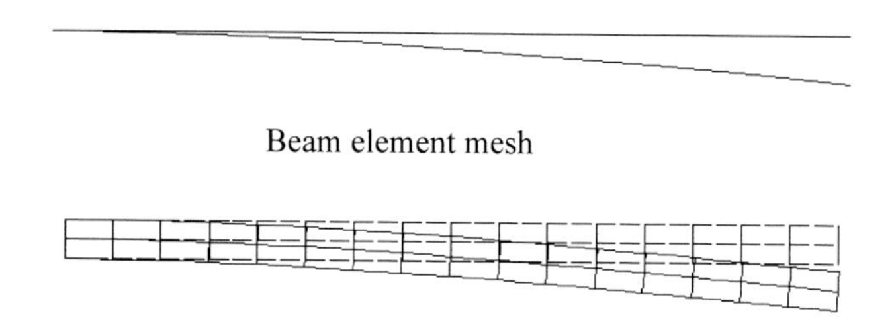

Fig. 10.11 Deformed shapes (solid lines) for the cantilever beam problem

10.3 EXAMPLE 3: NOTCHED PLATE

Problem definition

Figure 10.12 shows a plate of length *L*, width *W*, and thickness (in the *z*-direction) *t*, with a semi-circular notch of radius *R* and a circular hole of diameter *D* placed in the upper half, subjected to a uniaxial stress σ_o.

The numerical values used are: $W = 50$ mm, $L = 100$ mm, $R = 10$ mm, $D = 20$ mm, and $\sigma_o = 100$ MPa. The value of the thickness *t* is either 10, 100, or 200 mm. The objective of the analysis is to determine the stress concentration at the notch (point B) and at the hole (point A).

Geometry

There is no symmetry in this problem, hence the whole geometry must be modelled. The thickness in the *z*-direction, *t*, determines whether this problem can be modelled as 2D or 3D. For $t = 10$ mm, a 2D plane stress assumption would be applicable, whereas for $t = 200$ mm, 2D plane strain conditions are more appropriate. However, for $t = 100$ mm, the thickness is neither very small nor very large, hence a 3D assumption is more appropriate. However, the stress concentrations at the notch and around the hole are not affected by the magnitude of the thickness *t*. Therefore, for practical purposes, it is reasonable to use a 2D plane strain rather than a 3D assumption, even when $t = 100$ mm.

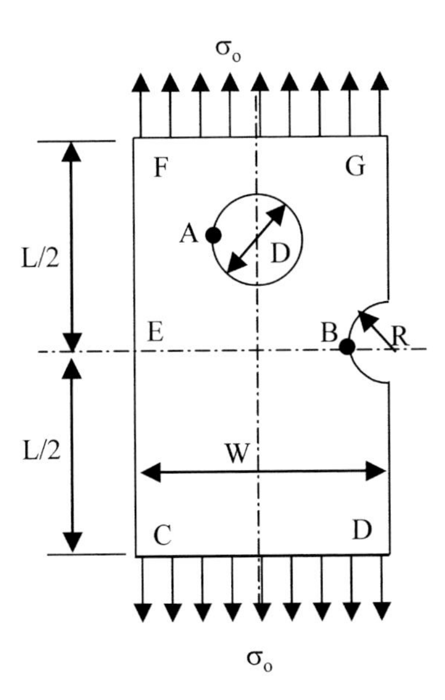

Fig. 10.12 Notched plate example

Material properties

Assuming an elastic analysis, the material properties needed are only Young's modulus (E) and Poisson's ratio (ν). The values used here are $E = 200$ GPa and $\nu = 0.3$.

Boundary conditions

Because there are no displacement constraints on any node, the whole plate would undergo rigid body motion, which invalidates the small deformation theory used in the FE formulation. Therefore, it is compulsory to introduce displacement constraints in all the Cartesian directions. This means that the FE model will not be exactly under the same boundary conditions as the real-life problem.

There are several options for imposing additional displacement constraints. One way of doing this is to fix the bottom surface, which will create high stresses at the edges (points C and D) but will not affect the overall accuracy at the high stress concentration areas in the notch and around the hole. Other displacement constraints are possible, provided they are not very close to the stress concentration regions. Figure 10.13 shows four possible displacement constraints that can be used to prevent rigid body motion.

Applied loads

A uniform tensile stress (distributed load) of magnitude σ_o is applied at the top surface for all four boundary conditions, as shown in Fig. 10.13. In BC(3) and BC(4), the stress σ_o is also applied at the bottom surface.

FE model

Figure 10.14 shows typical FE meshes that can be used to model this problem. The problem can be modelled as either 2D plane stress ($t = 10$ mm), 2D plane strain ($t = 200$ mm), or 3D ($t = 100$ mm).

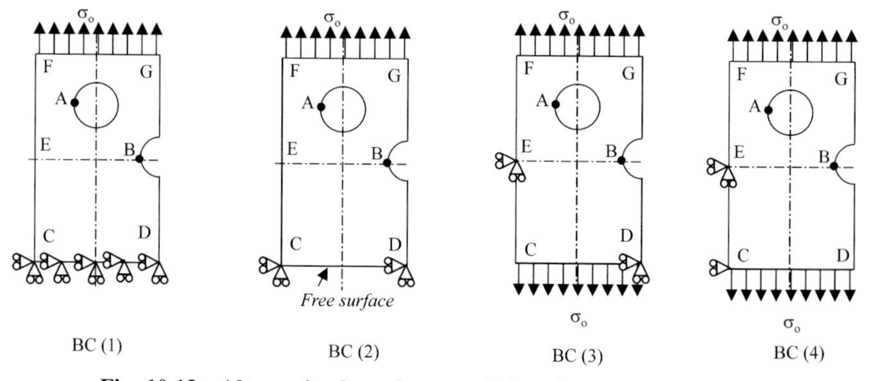

Fig. 10.13 Alternative boundary conditions for the notched plate

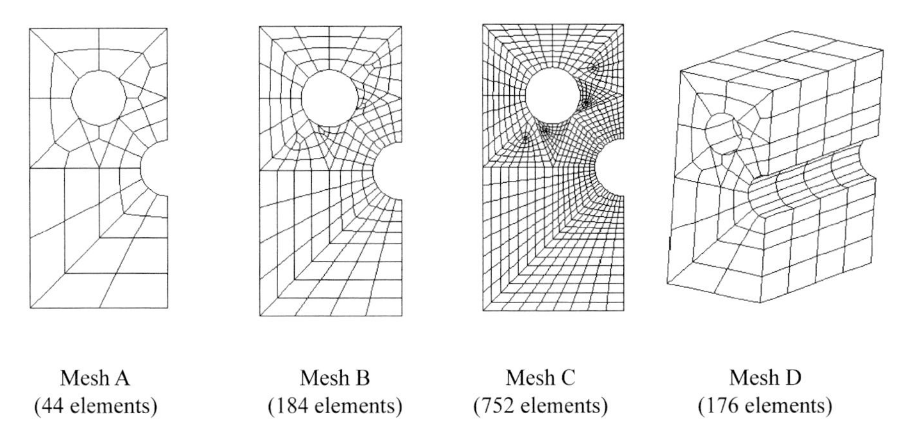

Fig. 10.14 Finite element meshes for the notched plate example

The deformed shapes for the four different boundary conditions are shown in Fig. 10.15 where it can be seen that the deformed shapes are slightly different because the locations of the constrained points are different. However, it is expected that the stress distribution around the hole and the notch will be similar for the four boundary conditions. This is illustrated in Table 10.1, which compares the stress concentration factors in the vertical direction for the different FE models. Therefore, it can be concluded that, if the objective of the analysis is to obtain the stress concentration at the hole and the notch, then using a 3D FE model is not necessary, as a 2D model can adequately represent the stresses.

The von Mises equivalent stress contours for the 2D plane stress elements are shown in Fig. 10.16. For all four boundary conditions, it can be observed

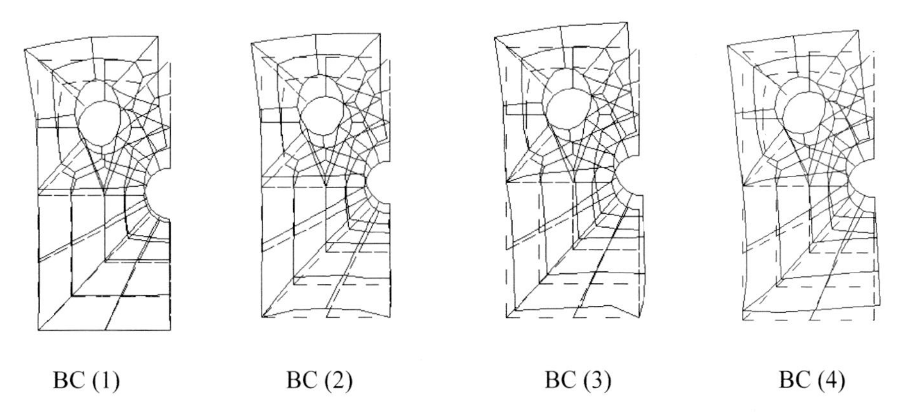

Fig. 10.15 Deformed shapes (solid lines) for the notched plate example (2D plane stress)

Table 10.1 Stress concentration for the notched bar example

Elements	*Boundary condition*	*Mesh*	*Stress concentration factor at point A (hole)*	*Stress concentration factor at point B (notch)*
2D plane stress	BC(1)	Mesh A	4.1505	4.5459
2D plane stress	BC(1)	Mesh B	4.1505	4.5459
2D plane stress	BC(1)	Mesh C	4.1605	4.5459
2D plane stress	BC(2)	Mesh A	4.1567	4.5465
2D plane stress	BC(3)	Mesh A	5.1445	4.2811
2D plane stress	BC(4)	Mesh A	4.4152	4.7078
2D plane strain	BC(1)	Mesh A	4.1505	4.5459
3D	BC(1)	Mesh D	4.1907	4.5465

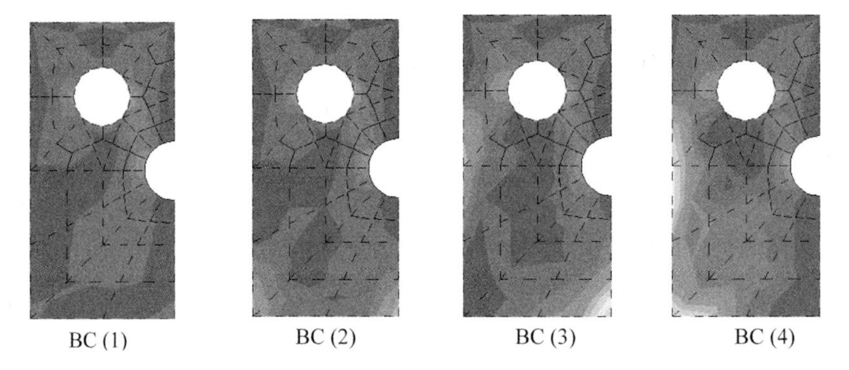

Fig. 10.16 von Mises equivalent stress for the notch plate example (2D plane stress)
(See colour plate section.)

that stress concentration exists around the constrained points, but the stress distribution around the hole and the notch are similar regardless of which boundary condition is used.

10.4 Example 4: Pressurized cylinder

This example concerns an open-ended cylinder of outer diameter D, wall thickness t, and axial length L, subjected to an internal pressure P, as shown in Fig. 10.17. The objective of the analysis is to determine the stress distribution through the wall thickness. The numerical values used are $D = 300$ mm, $t = 30$ mm, $L = 450$ mm, $P = 10$ MPa, with elastic material properties $E = 200$ GPa and $\nu = 0.3$.

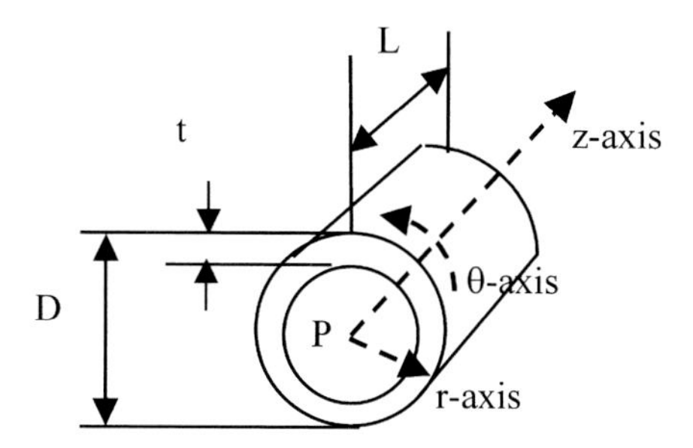

Fig. 10.17 Pressurized cylinder example

A number of alternative FE approaches can be used to model this problem, as follows:

(a) *2D plane strain model:* Because the length L is large compared to the thickness t, it is reasonable to assume that plane strain conditions are applicable. Therefore, any section perpendicular to the cylinder axis (the z-axis) can be modelled. Owing to symmetry, a quarter-model can be modelled with appropriate symmetry boundary conditions on lines AB and CD, as shown in Fig. 10.18(a). Note that since the behaviour of the cylinder is purely radial, it is also possible to model a smaller angular sector of less than 90°, for example, a 10° sector with appropriate symmetry conditions at lines AB and CD, as shown in Fig. 10.18(b). Note that the roller conditions on the inclined line CD are not aligned with the Cartesian axes, and may require the use of local axes in the FE software.

(b) *Axisymmetric model:* Because the pressurized cylinder is axisymmetric (in both geometry and loads), a section through the r–z plane along any angle θ can be used to represent the full problem, as shown in Fig. 10.19(a). Because the cylinder is long in the axial direction, and the deformation is purely radial, it is possible to model only a short axial length of the

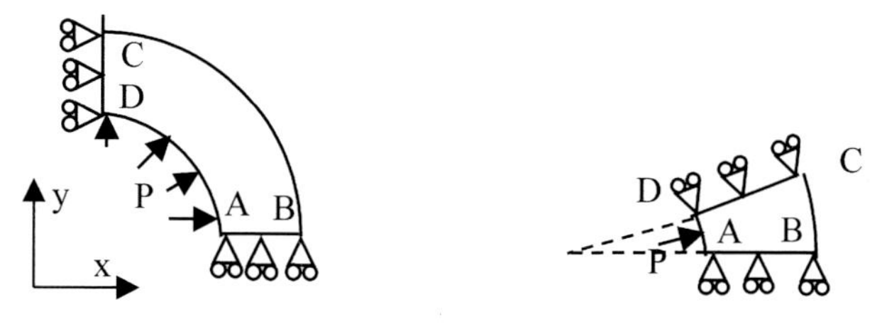

(a) Full geometry (symmetrical quarter) (b) Small angular sector

Fig. 10.18 Two-dimensional plane strain model of the pressurized cylinder example

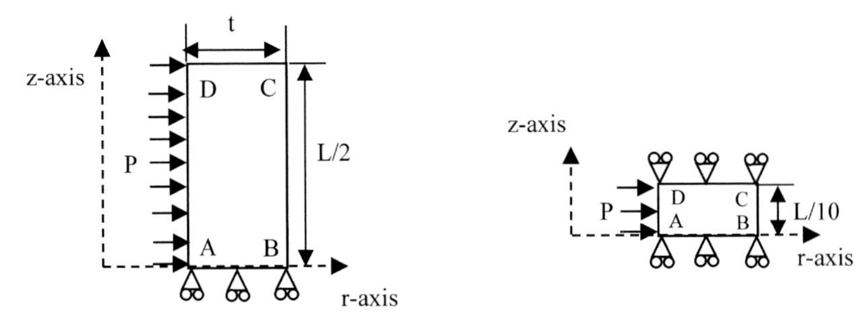

(a) Full cylinder length considered (b) Partial cylinder length considered

Fig. 10.19 Axisymmetric model of the pressurized cylinder example

cylinder with appropriate boundary conditions on the top and bottom surfaces, as shown in Fig. 10.19(b). To simulate a symmetrical half of the cylinder, symmetry roller conditions must be placed on the bottom surface. If a short axial length of the cylinder is used, roller conditions may also be placed on the top surface to simulate the larger length L. Note that there is no need to specify additional displacement constraints to prevent rigid body motion in the r-axis. This is because rigid body motion is automatically prevented in the r-direction by virtue of the axisymmetric FE formulation in which all the elements are 'ring' elements.

(c) *3D model:* It is possible to model the cylinder as a 3D problem, but this is not necessary because the stresses will be the same along the axial length of the cylinder. Figure 10.20 shows a symmetric eighth of a 3D model with the appropriate roller symmetry conditions.

(d) *Shell model:* Owing to the small thickness of the cylinder wall ($t/D = 0.1$), it is possible to use thin or thick shell elements, as shown in Fig. 10.21. Note that to simulate symmetry conditions, the displacements perpendicular to the symmetry planes and the slopes must be prescribed as zero.

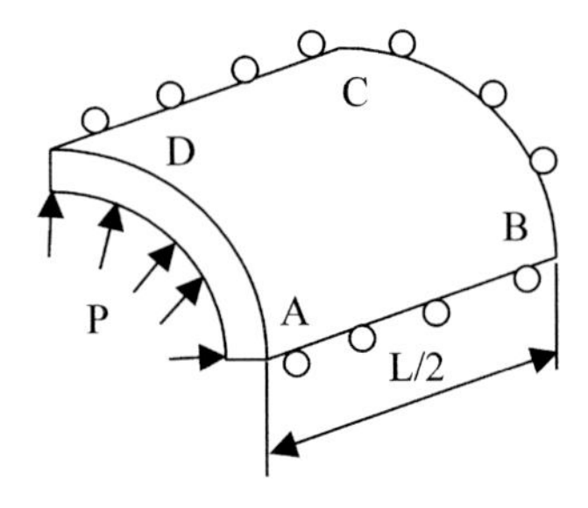

Fig. 10.20 Three-dimensional model of the pressurized cylinder example

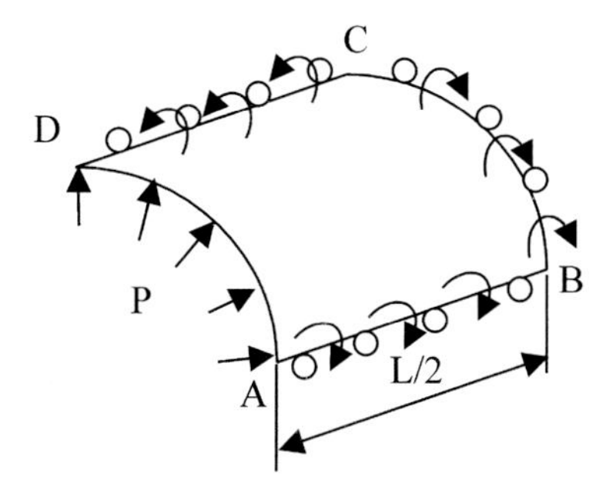

Fig. 10.21 Shell model of the pressurized cylinder example

Examples of suitable FE meshes that can be used to model the cylinder are shown in Fig. 10.22. From thick cylinder theory, the radial stress is expected to drop sharply from a peak at the inner surface. Hence mesh biasing should be used towards the inner surface. The deformed shapes are shown in Fig. 10.23.

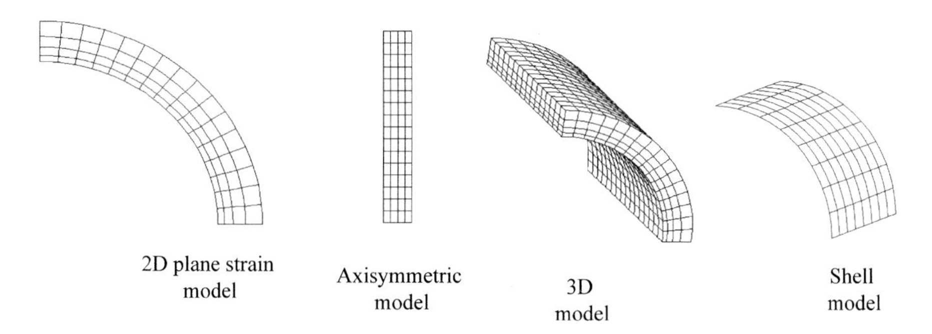

Fig. 10.22 Finite element meshes suitable for the pressurized cylinder example

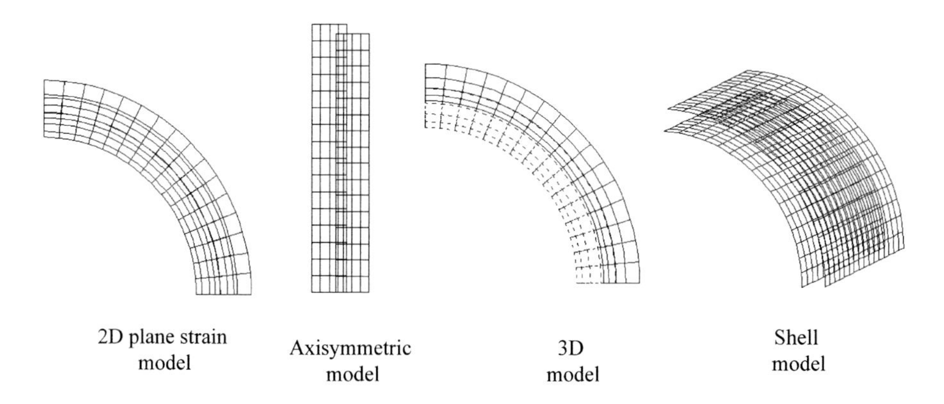

Fig. 10.23 Deformed shapes (solid lines) for the pressurized cylinder example

10.5 Example 5: Compressed Ring

Consider a ring of inner diameter D_1, outer diameter D_2, and thickness (in the z-direction) t_z, subjected to two compressive point forces, F, applied at opposite ends of the diameter, as shown in Fig. 10.24. The objective of the analysis is to determine the change in the diameter due to the compressive forces. The numerical values used are $D_1 = 90$ mm, $D_2 = 100$ mm, $t_z = 15$ mm, and $F = 1$ kN, with elastic material properties $E = 200$ GPa and $\nu = 0.3$.

Because the ring thickness (in the z-direction) is small, 2D plane stress conditions are appropriate. A symmetrical quarter of the ring with the appropriate symmetry boundary conditions is shown in Fig. 10.25(a). Note that the point force on the top of the quarter-model must be equal to $F/2$

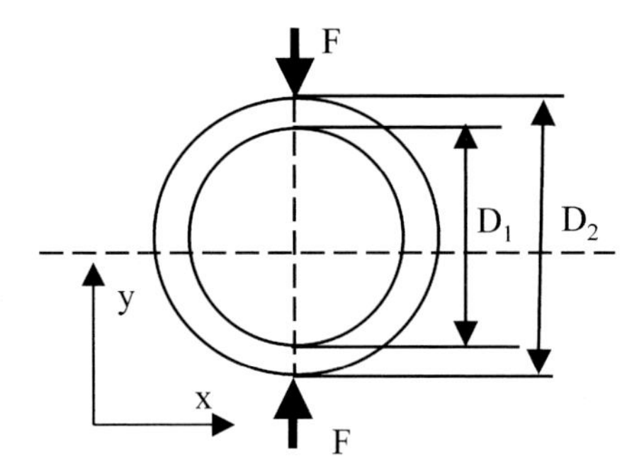

Fig. 10.24 Compressed ring example

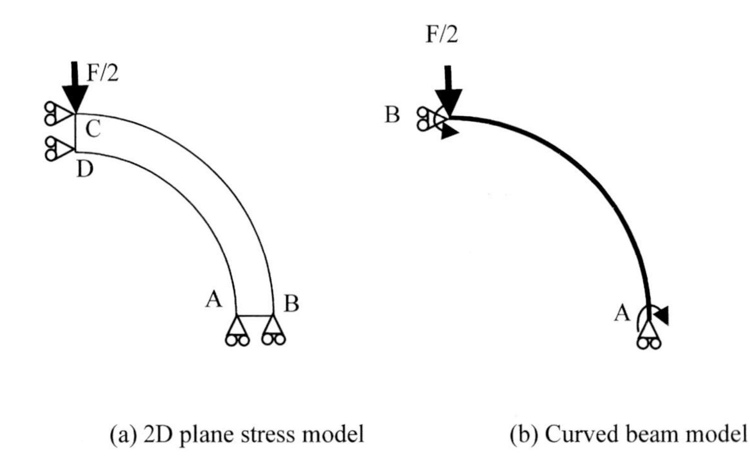

Fig. 10.25 Finite element models for the compressed ring example

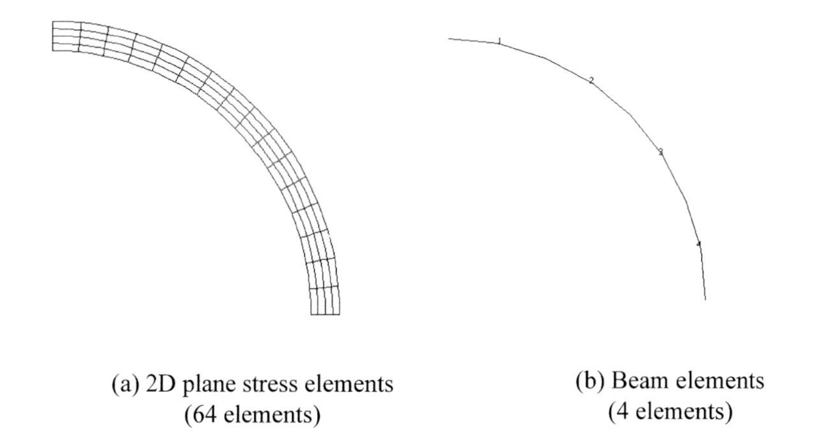

(a) 2D plane stress elements (64 elements)

(b) Beam elements (4 elements)

Fig. 10.26 Finite element meshes suitable for the compressed ring example

(not *F*), otherwise the quarter model will not represent the full model. It is also possible to use a curved beam model because the ring is thin. A symmetrical quarter beam model is shown in Fig. 10.25(b) with the appropriate symmetry conditions. Note that displacement, as well as slope, must be zero at points A and B to simulate the symmetry conditions.

Typical 2D plane stress and beam FE meshes are shown in Fig. 10.26. Note that stress concentration is expected around the point of application of the force. Hence mesh biasing should be specified around the point load. However, since the objective of the analysis is to determine the overall displacement, mesh refinement near the point of application of the force is not needed. The deformed shapes are shown in Fig. 10.27.

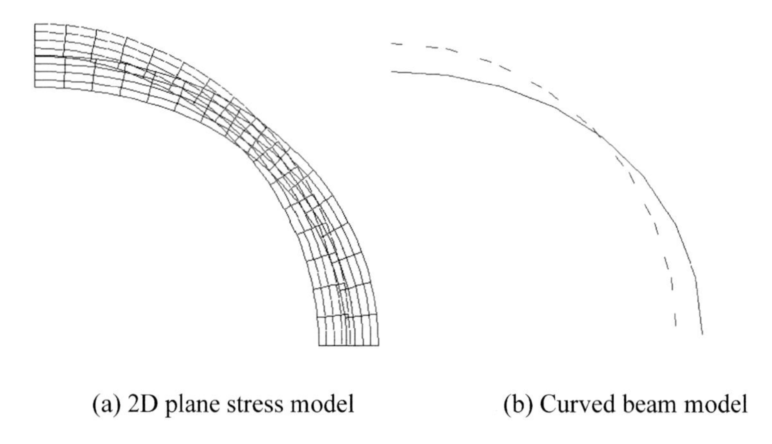

(a) 2D plane stress model

(b) Curved beam model

Fig. 10.27 Deformed shapes for the compressed ring example

10.6 Example 6: Pin-loaded plate

Consider a notched plate of width W, height H, and thickness L (in the z-direction), subjected to two equal forces of magnitude F applied through two rigid pins of diameter D, as shown in Fig. 10.28. The objective of the analysis is to obtain the stress concentration at the notch. The numerical values used are $W = 50$ mm, $H = 100$ mm, $L = 100$ mm, and $D = 4$ mm, with elastic material properties $E = 200$ GPa and $\nu = 0.3$.

The thickness of the plate in the z-direction is neither small nor large compared to the other dimensions, hence the model should be a 3D model. However, it would be much simpler to reduce the problem to a 2D geometry. Because the objective of the analysis is to analyse the stress concentration associated with the notch, then a 2D plane strain model would be acceptable.

A symmetrical half of the plate can be used as shown in Fig. 10.29(a), where displacement roller conditions are placed on the axis of symmetry (line AB). However, these displacement constraints are not sufficient to prevent rigid body motion in the x-direction. One way of avoiding rigid body motion is to fix a single point in the x-direction. It is convenient to apply an additional displacement constraint to point A that is far away from the notch area and the loaded pin.

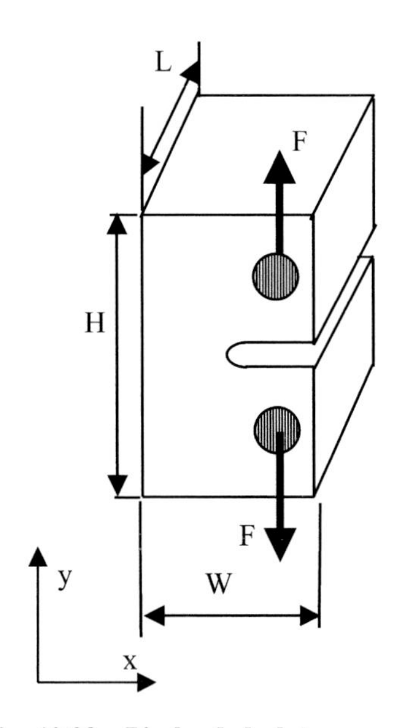

Fig. 10.28 Pin-loaded plate example

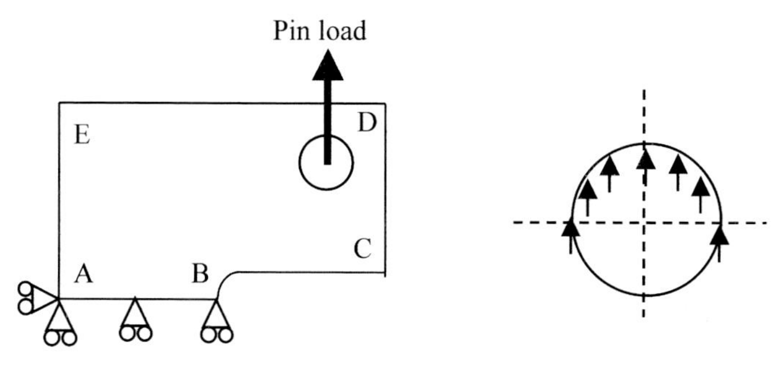

(a) Symmetrical half of the plate (b) Simulation of pin load

Fig. 10.29 Finite element models for the pin-loaded plate example

The pin load can be represented in a number of ways. Because the pin load is relatively far from the notch, any reasonable approximation can be used, provided the pin load is statically equivalent to the prescribed load. The simplest pin load model is to prescribe an upward point force on any point on the circular hole. A more accurate method is to prescribe a distributed load (compressive stress) on the points on the upper half of the circular hole, as

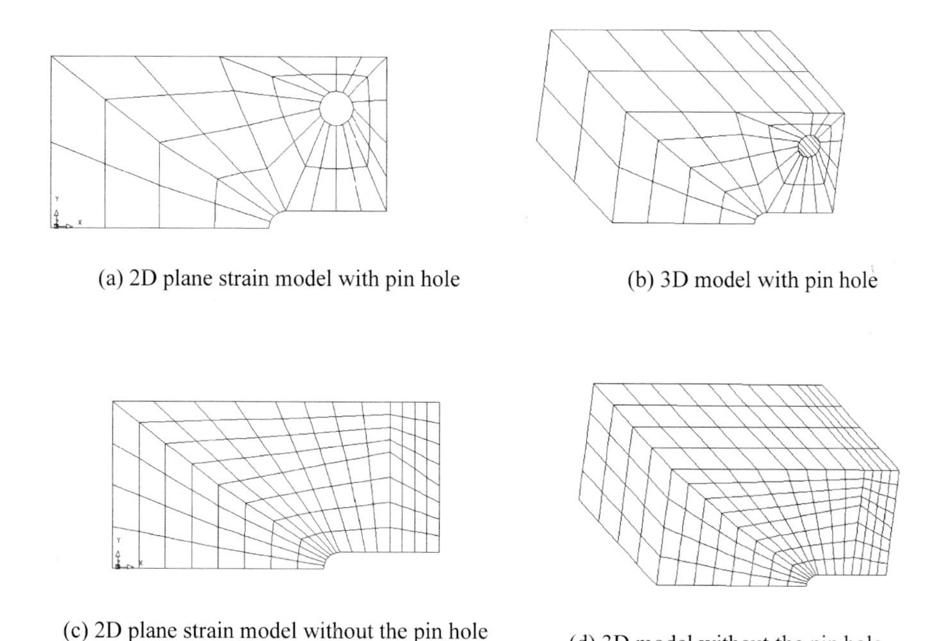

(a) 2D plane strain model with pin hole (b) 3D model with pin hole

(c) 2D plane strain model without the pin hole (d) 3D model without the pin hole

Fig. 10.30 Finite element meshes suitable for the pin-loaded plate example

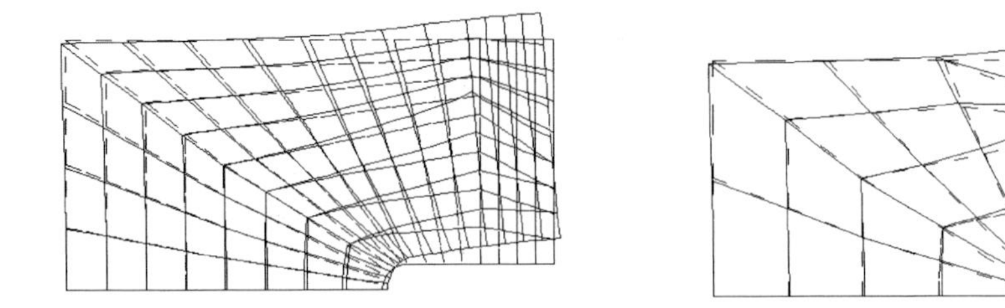

Fig. 10.31 Deformed shapes for the pin-loaded plate example

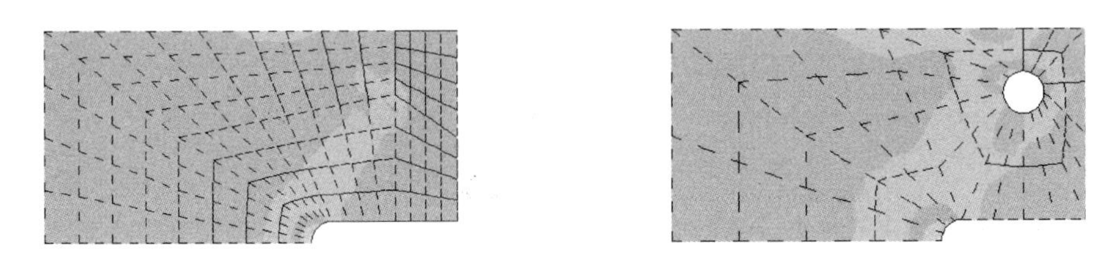

Fig. 10.32 von Mises equivalent stress for the pin-loaded plate example
(See colour plate section.)

shown in Fig. 10.29(b). The magnitude of the prescribed stress can be calculated by dividing the pin force by the surface area of the circular hole. A simpler alternative is to simply remove the pin hole and apply the point force to the node closest to the location corresponding to the centre of the pin.

Figure 10.30 shows typical 2D and 3D FE meshes suitable for modelling the plate – one incorporating the pin hole and one without the pin hole. The deformed shapes for both meshes, shown in Fig. 10.31, indicate that the overall deformation of the plate is similar. Similarly, the stress contours shown in Fig. 10.32 for both meshes show that although the stress concentration is different around the hole area, it is similar around the notch area.

10.7 SUMMARY OF KEY POINTS

- Symmetry should be used to reduce the size of the problem, with appropriate roller displacement constraints placed on the axes of symmetry.
- Quadratic elements are better at representing curved geometries and rapidly varying stresses.
- In problems where the body is under-restrained, the FE user must introduce additional displacement constraints, away from the region of interest, to prevent rigid body motion.

- If possible, the user should attempt to simplify the FE analysis, for example, by reducing a 3D problem to an equivalent 2D, beam, or shell problem.
- Some geometric features can be simplified provided that the overall effects on the deformations in the regions of interest are relatively unaffected.
- In beam, plate, or shell elements, the boundary conditions must cover slope as well as displacements.

References

(1) **Brebbia, C. A., Telles, J. C. F.,** and **Wrobel, L. C.** (1983) *Boundary element techniques*, Springer Verlag, Berlin.

(2) **Becker, A. A.** (1992) *The boundary element method in engineering*, McGraw-Hill, London.

(3) **Banerjee, P. K.** (1994) *The boundary element methods in engineering*, McGraw-Hill, New York.

(4) **Mitchell, A. R.** and **Griffiths, D. F.** (1980) *The finite difference method in partial differential equations*, Wiley, Chichester.

(5) **Fenner, R. T.** (1986) *Engineering elasticity*, Ellis Horwood, Chichester.

(6) **Boresi, A. P., Schmidt, R. J.,** and **Sidebottom, O. M.** (1993) *Advanced mechanics of materials*, Fifth Edition, John Wiley & Sons, New York.

(7) **Kreyszig, E.** (1979) *Advanced engineering mathematics*, John Wiley, New York.

(8) **Fenner, R. T.** (1989) *Mechanics of solids*, Blackwell Scientific Publications, Oxford.

(9) **Rees, D. W. A.** (1990) *Mechanics of solids and structures*, McGraw-Hill, London.

(10) **Crisfield, M. A.** (1991) *Non-linear finite element analysis of solids and structures*, Vol. 1, Wiley, Chichester.

(11) **Zienkiewicz, O. C.** and **Taylor, R. L.** (1991) *The finite element method*, Vol. 2, McGraw-Hill, London.

(12) **Hinton, E.** (1992) *Introduction to non-linear finite element analysis*, NAFEMS, Glasgow.

(13) **Becker, A. A** (2001) *Understanding non-linear finite element analysis*, NAFEMS, Glasgow.

(14) **Riks, E.** (1979) An incremental approach to the solution of snapping and buckling problems, *Int. J. Solids & Structures*, **15**, 529–551.

(15) **Crisfield, M. A.** (1980) Incremental/iterative solution procedures for non-linear structural analysis, *Numerical methods for non-linear problems*, edited by C. Taylor, E. Hinton, D. R. J. Owen, and E. Onate, pp. 261–290, Pineridge Press, Swansea.

(16) **Carslaw, H. S.** and **Jaeger, J. C.** (1959) *Conduction of heat in solids*, Oxford University Press, Oxford.

(17) **Timoshenko, S. P.** and **Goodier, J. N.** (1970) *Theory of elasticity*, Third Edition, McGraw-Hill, New York.

Bibliography

Bathe, K. J. (1982) *Finite element procedures in engineering analysis*, Prentice–Hall, New Jersey.

Belytschko, E., Liu, W. K., and **Moran, B.** (2000) *Nonlinear finite elements for continua and structures*, Wiley, Chichester.

Bittnar, Z. and **Sejnoha, J.** (1996) *Numerical methods in structural mechanics*, ASCE Press, New York.

Bonet, J. and **Wood, R. D.** (1997) *Non-linear continuum mechanics for finite element analysis*, Cambridge University Press, Cambridge.

Crisfield, M. A. (1997) *Non-linear finite element analysis of solids and structures*, Vol. 2, Wiley, Chichester.

Fagan, M. J. (1992) *Finite element analysis – Theory and practice*, Longman Scientific & Technical, London.

Hinton, E. and **Owen, D. R. J.** (1979) *An introduction to finite element computations*, Pineridge Press, Swansea.

Irons, B. and **Shrive, N.** (1983) *Finite element primer*, Ellis Horwood, Chichester.

Livesley, R. K. (1983) *Finite elements: An introduction for engineers*, Cambridge University Press, Cambridge.

Zienkiewicz, O. C. and **Taylor, R. L.** (1989) *The finite element method*, Vol. 1, McGraw-Hill, London.

* * * COMMUNICATION RESULT REPORT (NOV.30.1998 10:19AM) * * *

TTI A&A 5137315110

FILE	MODE	OPTION	ADDRESS (GROUP)	RESULT	PAGE
447	MEMORY TX		SUGAR BLDG	OK	P. 2/2

REASON FOR ERROR

E-1) HANG UP OR LINE FAIL
E-3) NO ANSWER

E-2) BUSY
E-4) NO FACSIMILE CONNECTION

Glossary of Terms

Aspect ratio of an element is the ratio of the largest to the smallest side of the element, used as a measure of the slenderness of the element. Large aspect ratios should be avoided.

Axisymmetric elements are elements with two degrees of freedom, the displacement in the radial (r) and axial (z) directions. The axial direction, z, is the axis of rotational symmetry about which the r–z plane is rotational through 360°. Axisymmetric elements can be regarded as 'ring' elements, despite being two-dimensional in appearance.

Banded matrix is a matrix with its nonzero coefficient placed in a 'band' around the diagonal. Special solvers can be used for banded matrices.

Bar elements are usually one-dimensional elements, which can only transmit axial forces, and have only one degree of freedom, the displacement along the bar.

Beam elements are elements that satisfy the beam bending equations, with two degrees of freedom (in two-dimensional problems): the axial displacement and the slope. The applied loads on beam elements are shear forces and bending moments.

Benchmarks are problems with known (and reliable) non-FE solutions that can be used to test the accuracy of FE programs.

Brick elements are three-dimensional elements (rectangular prism shape) with straight line or quadratic curved sides.

Cartesian coordinates are the orthogonal set of axes in the x, y, and z axes, also called global axes.

Compatibility equations are equations with differentials of the displacement functions, which must be satisfied to ensure that the displacements are continuous over the domain (with no overlap or gaps between adjacent elements).

Constitutive equations are the relationships between stress and strain incorporating material properties, such as Young's modulus and Poisson's ratio.

Contact problems are problems involving two domains that touch without necessarily being stuck together. Elements in the contact areas may stick or slip according to the prescribed coefficient of friction.

Continuity is used to describe the continuous nature of displacements across element boundaries, that is, no step changes in displacement between adjacent elements. Stresses, however, are allowed to be discontinuous across element boundaries.

Contour plots are colour plots showing lines of constant stress values within the domain. Plotted variables are extrapolated and interpolated to fall within the ranges of the contours.

Convergence of the solution means that as the FE mesh is refined the computed solutions approach the exact solution. In nonlinear problems, convergence is usually used to indicate the accuracy and reliability of the iterative procedures.

Deformed shapes are obtained by multiplying the nodal displacements by a suitable factor and adding them to the nodal coordinates. The result is an exaggerated plot of the deformed shape.

Degrees of freedom are the nodal variables, usually displacement components in the directions of the axes. Nondisplacement values, such as slopes at the nodes, are also used as degrees of freedom (e.g. in beam elements).

Energy formulation is used to derive the element stiffness matrix, utilizing the principle of minimum total potential energy.

Equilibrium equations are differential equations containing stresses. They are obtained by considering forces on a small differential volume.

Frontal solution is a special technique used to solve the FE algebraic equations by introducing elements, one at a time, to the solver. It is used to reduce the amount of computer storage required.

Functional is the function to be minimized in the variational or energy formulations. In a stress analysis problem, the functional is the total potential energy.

Gaussian elimination is a standard reliable technique for solving a system of linear algebraic equations.

Gaussian integration/quadrature is a well-established numerical algorithm used to perform numerical integrations. This technique is widely used in FE programs to calculate the integrals needed to obtain the element stiffness matrix and the element stresses.

Gaussian points are the total number of points used in the Gaussian quadrature technique. More points produce better accuracy, but consume larger computation time.

Galerkin method is a special form of the weighted residual technique widely used in FE formulations. Special weighting functions are used that are equal to the approximate (trial) solutions.

Geometric nonlinearity occurs usually in problems involving large deformations such as buckling problems.

Global axes are usually the x, y, and z (i.e. Cartesian) coordinates.

Hermitian elements are elements, such as beam elements, in which the degrees of freedom include variables such as slopes as well as displacements.

Hourglass mode is a special case of zero stiffness obtained for an element, which can be caused by reduced integration.

Hybrid elements are elements in which stresses as well as displacements are used as the independent variables. They are used in specialist applications such as the deformation of rubber-like materials.

Ill-conditioning of matrices may occur if the round-off errors become significant, that is, the equations solver becomes unstable. This can happen when rigid body motion takes place or when there are coefficients very different in magnitude in the same row, for example, when a very high value of Young's modulus is used for one material, but a very low one for another material.

Integration points are the Gaussian quadrature points used in the numerical integration.

Isoparametric elements are elements in which the same shape functions are used to describe the geometry and the displacements.

Iterations are an indirect way of solving nonlinear problems, based on successive corrections of an initial guess (trial solution).

Jacobian matrix is a matrix used to transform a set of variables from one set of coordinate axes (e.g. Cartesian components) to another set of coordinate axes (e.g. local components).

Lagrangian elements are elements in which the displacements are the only independent variables, such as truss and continuum elements.

Large deformations occur in problems involving nonlinear geometric behaviour, such as buckling problems, in which the deformations are so large that a new stiffness matrix has to be calculated.

Local axes are the axes used to describe the behaviour of each element without reference to the global axes. The Jacobian matrix is used to relate local variables to global variables.

Material nonlinearity occurs when the stress–strain relationships are not linear as in plastic or creep behaviour.

Mesh is usually used to refer to all the elements used to model a given problem.

Newton–Raphson method is a special method of solving nonlinear equations using an iterative approach based on either a constant or an updated slope in each iteration.

Nodes or *nodal points* are the points within the elements where the degrees of freedom (variables) are defined.

Patch test is a simple test of the potential performance of an element in which a state of constant strain is prescribed on a 'patch' of elements.

Pin-jointed structures are structures consisting of long straight line members connected by frictionless pin-joints that cannot support any bending moment.

Plane strain elements are used for two-dimensional 'thick' structures, where the normal strain is zero, that is, the x–y plane considered is very remote from the ends.

Plane stress elements are used for two-dimensional 'thin' structures, where the normal stress is zero, that is, the x–y plane is considered very thin.

Plate elements are used to model a plate under transverse loading, and can sustain membrane loads as well as bending moments.

Post-processor is the part of an FE program used to examine the computed values, usually in interactive graphics plots, such as deformed shapes and stress contour plots.

Pre-processor is the part of an FE program that deals with the generation of the data input, that is, mesh generation, boundary conditions, and load description.

Reduced integration is a less accurate Gaussian quadrature integration scheme in which the number of Gaussian points is reduced. Despite being a less accurate technique than full integration, reduced integration can improve the overall accuracy of several types of elements and consumes less computation time.

Rigid body motion is the motion of the body in any direction that generates no stresses. Rigid body motion must be prevented, otherwise the FE solutions become meaningless.

Ritz method is a technique used to solve partial differential equations by assuming a series trial solution that satisfies the boundary conditions of the problems.

Round-off errors are small errors in the computations caused by terminating long numbers after a given number of significant digits after the decimal points.

Shape functions are the interpolation functions that describe the behaviour of the geometry and/or the displacements within each element.

Shell elements are used to model thin or thick shell structures in which the membrane and bending actions are coupled.

Sparse matrix is a matrix with the majority of its coefficients equal to zero, such as the overall FE stiffness matrix.

Stiffness matrix is the matrix multiplying the displacement vector to equalize the external force vector. The stiffness matrix is symmetric and sparsely populated.

Symmetric matrix is a square matrix with its coefficients symmetrical about the main diagonal, that is, $a_{ij} = a_{ji}$. The FE stiffness matrix is a symmetric matrix.

Topology of an element is the order in which the nodes of the element are numbered, for example, clockwise or anticlockwise numbering.

Transpose of a matrix is obtained by interchanging its rows and columns.

Truss elements are used for pin-jointed structures in which only axial forces are transmitted.

Weighted residual method is a method of deriving the FE stiffness matrix by multiplying the partial differential equation by a suitable weighting function and minimizing the error.

Index